4-H Robotics: Engineering for Today and Tomorrow

Junk Drawer Robotics

Level 1

Give Robots a Hand

Presenter's Activity Guide

Junk Drawer Robotics Level 1
Give Robots a Hand

Table of Contents

Introduction to Junk Drawer Robotics 3

What Is 4-H Science? 4

Experiential Learning Process 5

Positive Youth Development 6

4-H Life Skills 7

What You Will Need for Junk Drawer Robotics 8

Focus for Give Robots a Hand 10

Overview of Module 1: Parts Is Parts 13

• Activity A – Think Like a Scientist 16

• Activity B – Communicate Like an Engineer 21

• Activity C – Build Like a Technician 25

• Activity D – Marshmallow Catapult Design Team 30

• Activity E – Marshmallow Catapult Build Team 33

Overview of Module 2: In Arm's Reach 35

• Activity F – Sense of Balance 38

• Activity G – ABC … XYZ 41

• Activity H – Arm in Arm Design Team 49

• Activity I – Arm in Arm Build Team 53

• Activity J – Pumped Up 55

• Activity K – Just Add Air Design Team 57

• Activity L – Just Add Air Build Team 59

Overview of Module 3: Get a Grip 61

• Activity M – Chopsticks 64

• Activity N – Just a Pinch 66

• Activity O – Hold On 68

• Activity P – One for the Gripper Design Team 70

• Activity Q – One for the Gripper Build Team 72

• Activity R – Twist of the Wrist Design Team 74

• Activity S – Twist of the Wrist Build Team 76

Glossary 78

References 79

Introduction to Junk Drawer Robotics

The goal of *4-H Junk Drawer Robotics* is to make science, engineering, and technology engaging and meaningful in the lives of young people. The activities do this by encouraging youth to use the processes and approaches of *science*; the planning and conceptual design of *engineering*; and the application of *technology* in their personal portfolios of skills and abilities.

The *Junk Drawer Robotics* curriculum is divided into three levels or books each around a central theme related to robotics design, use, construction, and control. Each level starts out with background information on working with youth, curriculum elements, and a focus on the concepts to be addressed.

- In Book 1, the theme is robotic arms, hands, and grippers.
- In Book 2, the theme is moving, power transfer, and locomotion.
- In Book 3, the theme is the connection between mechanical and electronic elements.

Modules within each book target major concepts of the theme. The modules each contain five or more activities that help the members develop an understanding of the concepts, create solutions to challenges, and develop skills in constructing alternatives.

Your role as a presenter in this curriculum is different from the normal role of the teacher that we may know from a school setting. It is not about mere transfer of knowledge from teacher to student. It is about assisting learners in developing their own knowledge and Problem solving skills. This is done by bringing together a scientific inquiry and engineering design approach to learning. Youth will learn about a topic by exploration. When given a problem, they will design a solution. Then, using what they have learned and designed, they will build or construct a working model.

The presenters of this curriculum may be teachers, after-school staff, or volunteers, including teens working with younger youth. In the case of teens, an adult coach or mentor could provide support, training, and guidance to teams of two or three teens who present the activities together. As a presenter, you will assist the youth in understanding the processes of science, engineering, and technology by how you ask questions and have them share their ideas, designs, and results.

The robotics curriculum is designed around three themes of science, engineering, and technology. Each module has activities in each of these three areas. As a framework, *4-H Junk Drawer Robotics* uses these simple definitions developed by Anne Mahacek, former 4-H teen member who is now a mechanical engineer and grad student in mechatronics.

 To Learn: **Science** is finding out how things work.

 To Do: **Engineering** is using what you found out to design something to work.

 To Make: **Technology** is using tools and processes to make something work.

What Is 4-H Science?

The National 4-H Science initiative addresses America's critical need for more scientists and engineers by engaging youth in activities and projects that combine nonformal education with design challenges and hands-on, inquiry-based learning in a positive youth development setting. These experiences engage youth and help them build knowledge, skills, and abilities in science, engineering, mathematics, and technology.

All 4-H Robotics activities and projects are:

- Based on National Science Education Standards and Standards for Technological Literacy,
- Focused on developing abilities in science, mathematics, engineering, and technology,
- Led using the Experiential Learning Model,
- Tied to developing Life Skills for youth, and
- Delivered in a positive youth development context by trained and caring adults.

National Science Education Standards (NSES)

The National Science Education Standards present a vision of a scientifically literate person and present criteria for our education system that will allow that vision to become reality (National Research Council, 1996). NSES outline what students should know, understand, and be able to do as they progress through their science education. Emphasis has shifted from being solely on "the content to be learned" to include "**how** students learn" and "**how** the content is presented."

Scientific Inquiry refers to the diverse ways in which scientists study the natural world and propose explanations of the world based on the evidence derived from their work (National Research Council, 1996). But scientific inquiry is not limited to the work of scientists. Young people driven by curiosity and given a structure can pose questions, make observations, analyze data, and

offer their own explanations. Supporting youth in developing the skills and understanding necessary to engage in scientific inquiry is a central focus of the NSES and 4-H Science.

Standards for Technological Literacy (STL)

Standards for Technological Literacy define what youth should know and be able to do in order to be technologically literate (International Technology Education Association, 2000). Technological literacy is important to all youth whatever path they pursue in life. They offer a common set of expectations for what students should learn in the study of technology and what is developmentally appropriate at different ages. The 20 standards address five areas: the nature of technology, technology in society, design, abilities for a technological world, and understanding the designed world. The activities in *Junk Drawer Robotics* help students develop literacy in each of these areas.

Science, Engineering and Technology Abilities

According to Horton, Gogolski, and Warkentien (2007), effective teaching in science, engineering and technology must focus on how youth learn the content **and** how the material is taught. Based on a review of the science, engineering, and technology education literature, the authors set forth 30 important processes and refer to them as Science, Engineering and Technology Abilities (Horton, Gogolski, & Warkentien, 2007). This set of science, engineering and technology life skill outcomes is emphasized throughout the 4-H Robotics curriculum. Examples of some of the 30 abilities that are developed in this curriculum include: observe, categorize, organize, infer, question, predict, evaluate, use tools, measure, test, redesign, collaborate, summarize, and compare. Each Module in *Junk Drawer Robotics* identifies particular abilities in science, mathematics, engineering, and technology that are focused on in that module.

Experiential Learning Process

Experiential learning allows young people to create or develop their own answer to a question instead of repeating "the answers" (Maxa, et. al., 2003). In the experiential learning process, youth are encouraged to think, explore, question, and develop decisions. Two important components of the experiential learning process are a period of reflection, during which the learner shares and processes the experience, and the application of new learning in "real world" situations. The experiential learning model contains five steps: experience (doing), share (what happened?), process (what's important), generalize (so what?), and apply (now what?) (Maxa, et. al., 2003).

The Experiential Model

This model helps leaders formulate activities to reflect the DO, REFLECT, APPLY in five steps.

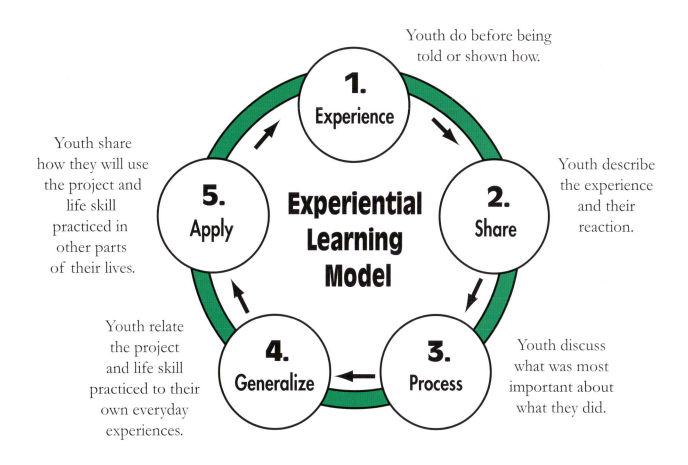

Youth do before being told or shown how.

1. Experience

Youth describe the experience and their reaction.

2. Share

Youth discuss what was most important about what they did.

3. Process

Youth relate the project and life skill practiced to their own everyday experiences.

4. Generalize

Youth share how they will use the project and life skill practiced in other parts of their lives.

5. Apply

Experiential Learning Model

Pfeiffer, J. W., & Jones, J. E., *Reference Guide to Handbooks and Annuals* 1983, John Wiley & Sons, Inc. Reprinted with permission from John Wiley and Sons, Inc.

Facilitation of the Experiential Learning Process

The key to the experiential learning process is that youth seek answers to questions rather than being given answers. This process requires facilitation instead of instruction. The role of adults in the experiential learning process is to facilitate the learning process, which means they become co-learners (Maxa, et. al., 2003).

Questioning Strategies

Questions suggested by the facilitator are designed to help promote discussion and engagement with life skills and science, engineering and technology abilities. The goal is to have the questions reside with the learner. Questions should promote discussion, interaction, and stimulate learner thinking. They should encourage ideas, speculation, and the formation of hypotheses. This type of questioning will not lead to a single right answer, but it will promote deeper understanding. It is important to allow adequate time for questions and discussions that engage youth and enhance learning. The process cannot be rushed.

Positive Youth Development

High quality 4-H Science programming provides valuable benefits in engaging youth. These programs also give youth the opportunity to engage in a positive youth development setting. Positive youth development occurs from an intentional process that promotes positive outcomes for young people by providing opportunities, choices, caring relationships, and the support necessary for youth to fully participate in families and communities. Simply, positive youth development seeks to develop young people as resources instead of problems to be managed (Lerner, 2005). Creating a positive youth development setting requires that youth are able to develop a sense of competence in social, academic, cognitive, health, and vocational aspects of life, a feeling of self-efficacy, a positive bond with a caring adult, a respect for societal and cultural norms, a sense of sympathy and empathy for others, and make contributions to self, family, communities, or society (Lerner, 2005).

4-H National Headquarters has identified four essential elements of positive youth development. They include:

- Belonging – To know they are cared about by others
- Mastery – To feel and believe they are capable and successful
- Independence – To know they are able to influence people and events
- Generosity – To practice helping others through their own generosity

To ensure that youth are engaged in a positive youth development setting, it is critical that learning includes the 4-H essential elements of positive youth development. This will give youth the opportunity to develop positive youth/adult relationships, practice life skills, and engage in the experiential learning model, which can promote mastery, independence, and generosity.

4-H Life Skills

Life skills are important in helping youth become self-directing, productive, contributing members of society (Maxa, et. al., 2003). 4-H programming strives to give youth developmentally appropriate opportunities to experience life skills, to practice them until they are mastered, and use these skills throughout a lifetime (Hendricks, 1998). The Targeting Life Skills model is based on the 4-H clover, which represents head, heart, hands, and health.

Targeting Life Skills Model

The "Targeting Life Skills Model" helps identify developmental life skills in 4-H and Youth Development Educational programs. They are grouped by:

Head Managing, Thinking
Heart Relating, Caring
Hands Giving, Working
Health Living, Being

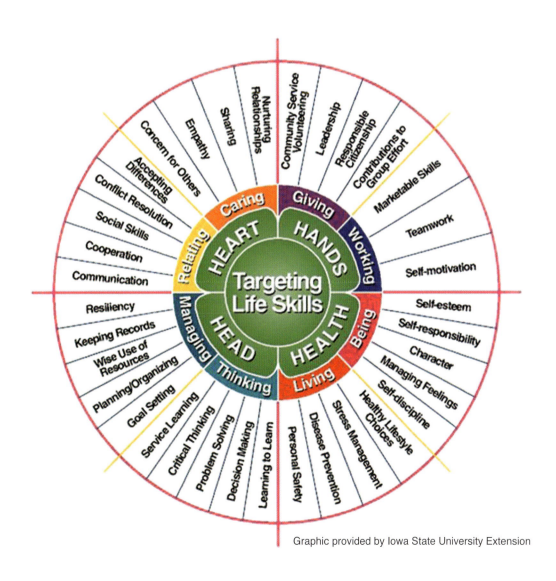

Graphic provided by Iowa State University Extension

What You Will Need for Junk Drawer Robotics

1. Trunk of Junk

The Trunk of Junk is a collection of items that will be used throughout this project. You may collect many of the items before you begin or add items as needed. Items you might collect are:

- Cardboard tubes from gift wrap, aluminum foil, paper towels, etc.
- Stationery supplies, like paper clips, binder clips, paper brads, etc.
- Clothespins, used household items, old toys, and items to take apart
- Coffee stirrers (sticks and straws), drinking straws, paper and plastic cups
- Construction paper, copy paper, graph paper, etc.
- Various pieces of cardboard; flats, boxes, tubes
- Assortment of small bolts, nuts, washers, and screws
- Wooden sticks (paint sticks, plant stakes, craft sticks, etc.) **Note**: Wooden sticks with holes are very useful in building. Holes can be predrilled or youth can drill holes as needed if you have appropriate tools. The website at *www.4-H.org/curriculum/robotics* includes tips on drilling holes in craft sticks.

2. Activity Supplies

Each activity has a list of the items to be prepared for that particular activity. Advance preparation or photocopying may be required. Some of the supplies may come from the Trunk of Junk, and some will be specific to the activities, but most will need to be organized before the meeting. Specific instructions are included in each activity.

3. Toolbox

This is a selection of basic hand tools that can be used throughout the project. These may be stored together in a toolbox, or collected before each meeting as needed. Commonly used tools include:

- Glue, tape, scissors (one or two per group)
- Low-temperature glue gun (two or three to share)
- Hand drill with different size bits
- Small hacksaw (to cut dowels and small boards)
- Pliers, wire cutters, screwdrivers, small hammer
- Bench hook or work surface to protect tables and chairs
- Paper towels, brush, cleanup supplies
- Punches (leather, craft, hole)

  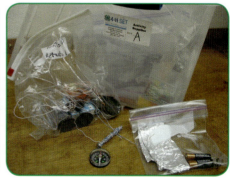

4. The Presenter's Activity Guide

This booklet is one of the three Presenter's Activity Guides: Give Robots a Hand, Robots on the Move, and Mechatronics. Each guide includes three to five modules made up of a series of activities. Three types of activities are within each module. In the To Learn activities, the focus is on gaining knowledge as a scientist. The To Do activities promote engineering design and innovation. And the To Make activities develop technological skills.

5. Robotics Notebook

There is one Robotics Notebook for the three levels of the *Junk Drawer Robotics* curriculum. Using the notebook encourages members to think like scientists and engineers. In their notebook, they will record their ideas, collect data, draw designs, and reflect on their experiences. It also provides specific information for the challenges. Each youth should have his or her own Robotics Notebook. If this is not possible, a blank notebook can be substituted. Graph paper will work best. Youth will have to record both the questions and their responses in a blank notebook, and leaders will have to make copies of supporting material.

6. Website and On-line Resources

The 4-H Robotics website, *www.4-H.org/ curriculum/robotics*, provides supporting information, including background about robotics concepts, tips, and resources for leaders. You will find detailed tips on gathering tools, locating supplies, and instructions on how to make some of your own materials for *Junk Drawer Robotics*. There are also resources specifically for coaches or mentors working with a team of teens as they present the *Junk Drawer Robotics* modules. The role of a coach or mentor is to provide the teens with back-up support and resources. The website also has an overview of how the *Junk Drawer Robotics* activities are related to other parts of *4-H Robotics: Engineering for Today and Tomorrow* curriculum and suggestions for implementing the curricula in different settings like camps, events, clubs, or afterschool programs.

7. Additional 4-H Robotics Opportunities

Extend your experience with robotics with another part of the *4-H Robotics: Engineering for Today and Tomorrow* curriculum. *4-H Virtual Robotics* provides the opportunity to utilize an interactive computer game environment to learn about the science and engineering of robots. *4-H Robotics Platforms* provides a curriculum to use as you explore a robotics kit and learn to design, build, and program a robot. For more information or to order the curriculum, visit the website at *www.4-H.org/curriculum/robotics*.

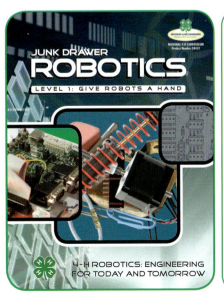
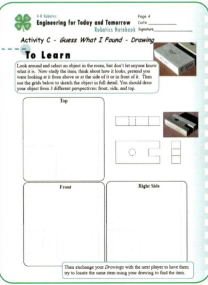

Focus for Give Robots a Hand

Introduction

In this level of Junk Drawer Robotics, we will be exploring and learning about robot arms. The arm is one of the most useful and imitated parts of the human body. It is flexible with joints — shoulder, elbow, wrist — that allow it to be placed in many positions to reach and grab. The hand with its fingers and thumb can grab, hold, pick up, and more.

Designs for robot arms can allow a robot to be used to grab, lift, move, or position items into a machine, to control a spot welder, or to assemble parts for an item.

In this level, Give Robots a Hand, we'll explore, design, and build while learning about many of the Big Ideas in the next column. They include power sources, arm designs, and assembly and more. We hope you will enjoy helping young people learn and grow during this process.

Big Ideas

Scientific habits of mind (observation, computation, communication, evaluation) are an important element of science literacy.

Form and function are essential considerations of quality design.

Engineering design is a purposeful process of generating and evaluating ideas that seeks to develop and implement the best possible solution to a given problem.

Robots often use simple machines (such as levers) as tools to accomplish their intended function.

The physical design (form) of a robot is based on its intended function, and it is often necessary to choose certain design elements (e.g. pneumatics, levers, etc.) or otherwise constrain the design in order to achieve the goal.

Understanding of underlying physical science and mathematics concepts is necessary in making engineering design decisions.

Connected Ideas

Robot Mobility is presented in
Virtual Robotics: Robot Tractor Pull,
Junk Drawer Robotics: Get Things Rolling,
Junk Drawer Robotics: Get a Move On,
Robotic Platforms: Build a Bot, and
Robotic Platforms: Pick a Motion Challenge.

Electronic Systems/Circuits is presented in
Virtual Robots: Electronics: Power Up!,
Junk Drawer Robotics: Watt's Up?,
Junk Drawer Robotics: Robots on the Move, and
Junk Drawer Robotics: It's Logical.

Robot Mechanics/Simple Machines is presented in
Virtual Robotics: Robot Mechanics,
Virtual Robotics: Robot Tractor Pull,
Junk Drawer Robotics: Give Robots a Hand,
Junk Drawer Robotics: Get a Grip, and
Robotic Platforms: Gears and Levers.

National Science Education Standards (NSE)

Unifying Concepts and Processes:
Form and function

Science as Inquiry:
Abilities necessary to do scientific inquiry

Physical Science Standards:
Motion and forces
Light, heat, electricity, and magnetism

Science and Technology:
Abilities of technological design
Understanding about science and technology

Standards for Technological Literacy (STL)

The Nature of Technology
Scope of technology
Core concepts of technology
Relationships among technologies

Technology and Society
Influence of technology
Use of technology

Design
Attributes of design
Engineering design
Problem solving

Abilities for a Technological World
Apply the design process
Use technological systems
Impact of products and systems

The Designed World
Energy and power technologies
Transportation technologies
Manufacturing technologies

Science, Engineering and Technology Abilities Developed in this Level

- Build/Construct
- Communicate/Demonstrate
- Collect Data
- Design Solutions
- Draw/Design
- Hypothesize
- Observe
- Predict
- Redesign
- Test
- Use Tools

Life Skills Practiced in this Level

- Communication
- Contributions to Group Effort
- Critical Thinking
- Keeping Records
- Problem Solving
- Sharing
- Teamwork

Module 1: Parts Is Parts

Overview of Activities in this Module

To Learn
Activity A — Think Like a Scientist
Activity B — Communicate Like an Engineer
Activity C — Build Like a Technician

To Do
Activity D — Marshmallow Catapult Design Team

To Make
Activity E — Marshmallow Catapult Build Team

Note to Leader

When we look at how things are made, we come across many words that relate to the making of things. Items can be made, constructed, built, manufactured, produced, or fabricated.

The two primary aspects of how things are made are the design elements and the manufacturing elements. Design elements include the geometric shape/size (round, square, long, wheels) of the item's parts. We can find many basic shapes in items, such as arcs, triangles, and columns. Manufacturing elements include the sequence of processing steps (forged, drilled, cast, cut, fabricated), and how the raw materials were used, modified, or changed in the assembly of the items.

Similar to the design and manufacturing of an item is the concept of form and function. Form is how something looks or feels. An item may be made a certain way so it looks good, has a great color, or is pleasing to hold — this is part of its form. The function is how something works or accomplishes a desired task. An item may be made to hold something, move something, or entertain us — this is part of its function.

There are different ways of thinking about the relationship between form and function. One viewpoint is based on Louis Sullivan's famous axiom, "Form follows function." This basic rule for design means that if an object has to perform a specific function, its design must first support that function. An example of this is the top-down design process, based on a challenge, need, or problem to be solved.

Another point of view is from the artist, who looks first at form and second at the item's function. This point of view gives rise to a bottom-up design process, given specific tools, materials, or supplies to accomplish the task. Both views can come into play in good design.

The architect Frank Lloyd Wright made this statement regarding form and function: "Form follows function — that has been misunderstood. Form and function should be one."

Raymond Loewy, referred to as the Father of Industrial Design, explained form and function this way: "I once said that the most difficult things to design are the simplest. For instance, to improve the form of a scalpel or a needle is extremely difficult, if not impossible. To improve the appearance of a threshing machine is easy. There are so many components on which one can work."

And, he said, "It would seem that more than function itself, simplicity is the deciding factor in the aesthetic equation. One might call the process, beauty through function and simplification."

When beginning to design, form and function must be considered. Both are necessary for good products; how they are achieved is up to the designer.

What you will need for Module 1: Parts Is Parts

- Robotics Notebook for each youth
- Trunk of Junk, see page 8
- Activity Supplies
  - Cardboard or pegboard work base; rectangle about 6 inches by 12 inches, at least one per group
  - Paper clips, binder clips in different sizes, about 10-15 per group
  - Full boxes of paper clips (at least one per group)
  - Paper brads, about 5-10 per group
  - Clothespins or other fasteners about 5-10 per group
  - Coffee stirrers, about 10-15 per group
  - Straws, about two to six per participant
  - Paint sticks (some with holes), about 5-10 per group
  - Craft sticks (some with holes), about 10-15 per group
  - Washers, about 5-10 per group
- Posters, Handouts, and other items (Note: Most of these elements are in the Robotics Notebook, but you may choose to create an additional poster or handout especially if youth do not have a copy of the Robotics Notebook.)
  - Optional: Copies of tables and graphs on pages 19-20
  - Optional: Poster/Handout of Manufacturing Processes on page 27
  - Optional: Poster/Handout of Design Shapes on page 28
  - Optional: Poster/Handout of Marshmallow Catapult Challenge on page 32
- Toolbox
  - Glue
  - Tape
  - Scissors, 1 or 2 per group
  - Low-temperature glue gun (two or three to share)
  - Hand drill with bits
  - Small hacksaw (to cut dowels and small boards)

Timeline for Module 1: Parts Is Parts

Activity A — Think Like a Scientist

- Activity will take approximately 20 minutes.
- Divide youth into small groups of two or three.
- For each team, arrange a random selection of parts on a table or work space.

Activity B — Communicate Like an Engineer

- Activity will take approximately 20 minutes.
- Divide youth into groups of four.

Activity C — Build Like a Technician

- Activity will take approximately 20 minutes.
- Divide youth into teams of two or three.

Activity D — Marshmallow Catapult Design Team

- Activity will take approximately 20 minutes.
- Divide youth into groups of two or three.

Activity E — Marshmallow Catapult Build Team

- Activity will take approximately 30 minutes.
- Use the same groups from Activity D, Marshmallow Catapult Design Team.
- The Marshmallow Catapult can be adapted into a balance beam scale to be used in Activity F, Sense of Balance.

Big Ideas

- Science habits of mind
- Form and function
- Engineering design process

NSE Standards

- Systems, order, and organization
- Form and function

STL

- Core concepts of technology
- Relationships and connections
- Attributes of design
- Manufacturing technologies

SET Abilities

- Categorize/Order/Classify
- Compare/Contrast
- Communicate/Demonstrate
- Draw/Design
- Build/Construct

Life Skills

- Keeping Records
- Critical Thinking
- Communication

Performance Tasks For Youth

You will learn the importance of identification as you make observations and sort materials based on selected attributes. You also will record data into charts and graphs.

You will describe an object by drawing and writing a description of it. You also will have to determine items that others have described.

You will use the engineering design process to complete a building challenge that involves using manufacturing processes and design shapes.

You will plan and design a swinging arm trebuchet-style catapult to launch marshmallows.

You will assemble parts, use simple tools, make modifications, and record information in your Robotics Notebook as you build a catapult.

Success Indicators

Youth will be able to select and categorize items based on observation and comparison of common attributes. Youth will be able to create their own charts and graphs based on data gathered.

Youth will be able to utilize communication skills to describe an object using sketching and drawing techniques as well as written and verbal descriptions.

Youth will be able to use the engineering design process in completing a challenge that includes using basic design shapes and tools.

Youth will be able to record thoughts, ideas, and design plans for a trebuchet-type catapult.

Youth will be able to use tools and parts to build a catapult of their design.

Activity A – Think Like a Scientist

Performance Task For Youth

You will learn the importance of identification as you make observations and sort materials based on selected attributes. You also will record data into charts and graphs.

Success Indicators

Youth will be able to select and categorize items based on observation and comparison of common attributes. Youth will be able to create their own charts and graph based on data gathered.

List of Materials Needed

- Robotics Notebook
- Collection of parts – Each team should have 15 to 30 items to sort. Teams do not have to have the same number of items, just a selection of different items to sort. Try to have different sizes and colors of paper clips; binder clips; clothespins; craft, paint, or wood sticks; paper brads; coffee stirrers; and drinking straws. If you can, try to use items that you already have available and that you'll be using in the builds.
- Optional: Copies of tables and graphs on pages 19-20

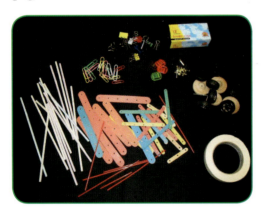

Activity Timeline and Getting Ready

- Activity will take approximately 20 minutes.
- Divide youth into groups of two or three.
- For each team, arrange a random selection of parts on a table or work area.

Sorting 1

1. Ask participants to look at and observe the items. Then ask the youth to sort the objects by color and shape, e.g., blue circle, red rectangle. Youth will then fill out the chart and graph in their notebook. In the notebook, youth should keep count of the different categories they've created and used.

2. Discuss with the youth the importance of keeping data in their notebooks, and how important it is to use charts and graphs. The youth should understand that their notebooks should be neat and easy for anyone to read and to understand what will happen during the experiment. For this activity, it means that anyone reading the notebook will be able to gather the same materials and have similar categories and groupings.

Sorting 2

3. Next, ask the participants to think about other ways they could group or sort the items. Have others sort the items into new groups using a different sorting criterion. Let the youth determine how they will sort the items, such as big/little, round/not round, flat/raised, rough/smooth, size, shape, or material. Have the youth sort the items, but not tell the other members of the entire group how they sorted. Have those who are not members of the group try to guess or predict how the items are sorted by observing the sorted items.

4. Youth should count the number of items in each new sorted grouping. Then they should

create their own chart and graph headings in their notebooks and record the numbers in their notebook.

5. Discussion: After the members have guessed the grouping, ask the participants to share other ideas on ways to sort the parts or materials.

6. Then use the *Note to Leader* information on page 13 to share about form and function.

Sorting 3

7. After the discussion, again ask youth to observe the parts carefully. Encourage them to notice how the parts are designed and how they are to be used.

8. Ask participants to sort the parts (items) according to:

   a. The item's use (**the function** — hold things, provide shelter, move stuff, etc.)

   b. The item's shape, appearance, or configuration (**the form** — the look of the items, colors, shapes, etc.)

9. Discussion: Have youth record their findings in their Robotics Notebook. Youth should create

a chart that lists the item, the function of the item, and the form of the item.

10. Have the youth share their ideas for form and function with the rest of the group. There are different possibilities for form and function; people use different objects in different ways. There are no wrong answers, only different opinions.

Example 1. All the items in the photo below are used to hold or fasten items together – they all have the same function. But some are pointed, some springy, some wrap around – they have different forms.

Example 2. Is the color (red) part of the item's *form* or *function*? Some might say red is just the *form*. The item can perform its function (hold paper together) if it is red, green, yellow, or a different color. But others may say that red is a *function* that might be used to help sort or group things based on the color. Files on robots are red, files on cars are yellow, files on gardens are green, etc.

Is it form or is it function?

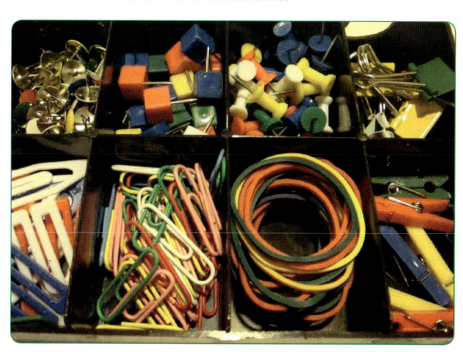

Sharing and Processing

As the facilitator, help guide youth as they question, share, and compare their observations. Before they share with the group, have youth reflect on the activity in their Robotics Notebook. You may choose one of the questions below as a prompt. If necessary, use more targeted questions as prompts to get to particular points. Remember these questions are not about getting one right answer.

- How did you select your sorting method?
- What knowledge did you need in this sorting exercise?
- Why could it be important to sort and classify parts?
- Why are the items we sorted made the way they are made? (Is it for the function of the part or is it the look and feel of the part?)
- How does classifying help determine how parts are made or used?

Generalizing and Applying

- What are some other things that are grouped or classified in science?
- What are the form and function of some other items?
- Youth can be asked to apply sorting to things in their room or other rooms in their home, at school, or in other locations. They can share their experience with the group.
- Ask the youth to determine the form and function of other items they have at home or see around town.
- Youth also can apply what they have learned in Activity B.

 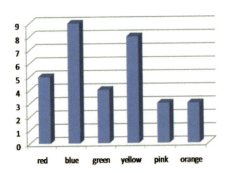 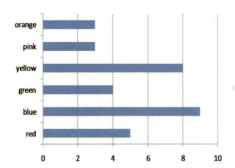

Sample graphs

Sorting 1 – Color and Shapes

Color / Shape	Square	Triangle	Circle	Rectangle	_____	_____	Total
Red							
Blue							
Green							
Yellow							
_____							
_____							

Graph 1 – Number by Color

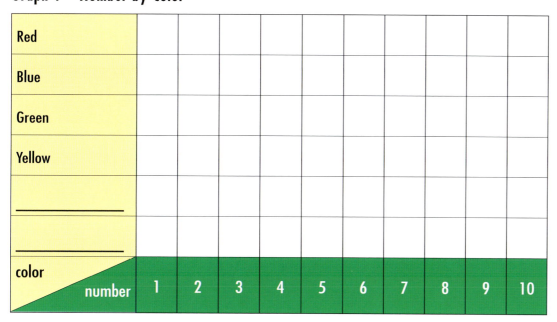

color										
Red										
Blue										
Green										
Yellow										
_____										
_____										
number	1	2	3	4	5	6	7	8	9	10

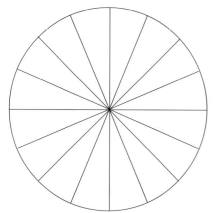

Pie Graph 1A - By Color

Sorting 2 – _____

	_____	_____	_____	_____	_____	_____	Total
_____							
_____							
_____							
_____							
_____							

Graph 2 – _____

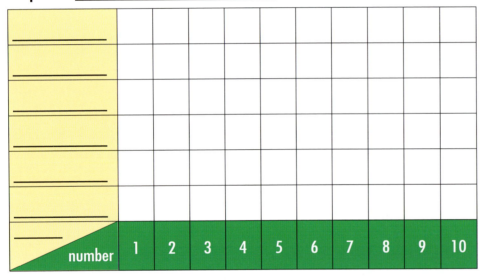

_____										
_____										
_____										
_____										
_____										
_____										
_____										
number	1	2	3	4	5	6	7	8	9	10

Sorting 3 – Form and Function

Item	What is its Function?	What is its Form?

Activity B - Communicate Like an Engineer

Performance Task For Youth

You will describe an object by drawing and writing a description of it. You also will have to determine items that others have described.

Success Indicators

Youth will be able to utilize communication skills to describe an object using sketching and drawing techniques as well as written and verbal descriptions.

List of Materials Needed

- Robotics Notebook
- Handouts/blank paper for drawing and writing activities

Activity Timeline and Getting Ready

- Activity will take approximately 20 minutes.
- Divide youth into groups of four.
- The Communicate Like an Engineer activity is played as a combination of two games. One is "I looked around and guess what I found," in which others have to guess the item they selected. The second game is "Telephone" where one person whispers a phrase to the person next to him or her and the phrase is repeated to the next person and so on until the person at the end indicates what he or she heard.
- In the Communicate Like an Engineer game, each youth will look around the area and select an item to be guessed, and then will use different communication types to try to have others identify the item. Each round will use a different communication skill as the notebooks are rotated to the next member.

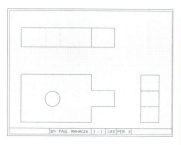
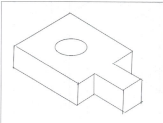

- Rectangular
- Wooden
- Has a hole through it in the large surface
- Is about 1 inch thick, 3 inches wide, and 5 inches long
- Light tan color
- Smooth surfaces
- One end has a tang

Note

We are asking the youth to draw and write descriptive narrations. These are not easy tasks and the reason we are doing them is to try to build knowledge, skills, and practice in doing them.

During the discussions, be careful not to have the quality of the drawings or writing be ridiculed but rather focus on the type of communication that will help scientists or engineers share their work and testing.

We should find that it is a combination of communicating styles that will help the most, prompting us all to become better drawers, writers, sketchers, chart makers, and number users.

Experiencing

1. Read with the students the Career Connection that appears at the end of Activity B and in the Robotics Notebook. Throughout these activities, students will use their Robotics Notebook as a tool to help them organize and record their scientific thinking, their engineering designs, and their building with technology.

   a. Discuss how their notebook can be used to help communicate their ideas. Engineers and scientists use notebooks to record notes and their work for themselves and to verify with others what they have tried, and what worked and didn't work.

   b. All of the notes are important and valuable. The notes on what didn't work are as important as the notes on what did work.

   c. An engineer's notebook that has been well kept, dated, and signed by the inventor can be the difference in being the first to get credit or a patent on an idea or design.

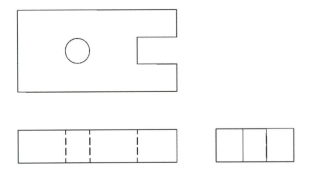

Round 1

2. Organize youth into groups of four. Each person will find an object in the room and draw it in their Robotics Notebook. The object should be drawn in 2-D orthographic projection, a drafting/drawing method in which two dimensions of an object are drawn as viewed from different directions. The object is drawn as it is seen from the front, side, and top views.

Round 2

3. Have youth rotate their Robotics Notebook to the next person in the group. The youth will look at the drawing in the notebook and try to determine the item drawn, but they should **not share** what it is with others. Then, on the next page in the Robotics Notebook, they should describe in words the object that was drawn. Try to encourage the use of adjectives to describe the item. Use shapes, possible materials, and its form. Only use words that describe the object; do not name the object in the description, nor describe the object's function (e.g., it shows the time).

- Rectangular
- Wooden
- Has a hole through it in the large surface
- Is about 1 inch thick, 3 inches wide, and 5 inches long
- Light tan color
- Smooth surfaces
- One end has a tang

Round 3

4. Rotate the Robotics Notebooks again and have the next youth guess what the object is by only using the description of words from Round 2. Again, do not share what the item is with anyone else. Then have this member draw a 3-D picture-style image of the object. The grid lines should help draw in a 3-D isometric style.

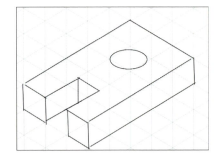

Round 4

5. Rotate the notebooks to the fourth member of the group. Have them guess what the item is from the 3-D drawing. Again, do not share

yet what the item is with anyone else. If there is more than one four-person group, take turns describing the item outloud to see if other groups can guess the item. Then have the original drawer verify the selected items.

Round 5

6. Return the notebook back to the original owner. Have youth discuss which type of drawings and which type of descriptions were easiest to use and to understand.

7. Discuss each type of communication, its limitations, and how it affected the way youth tried to guess the object. Use science-related questions that promote inquiry. Youth should record notes in their notebooks. Remind them to date and sign the notebooks to verify their notes.

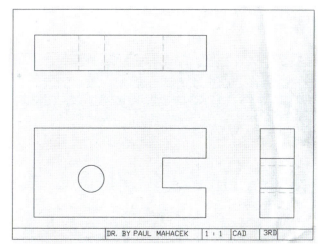

DR. BY PAUL MAHACEK | 1 : 1 | CAD | 3RD

Sharing and Processing

Have participants record their experiences in their Robotics Notebook. They should record the things they learned about communications related to note taking, drawing, and picture graphics:

- How do we communicate designs?
- What descriptions and drawings were harder or easier to understand?
- Which communications method was easier to perform?
  Round 1: 2-D Orthographic Projection
  Round 2: Written Description
  Round 3: 3-D Perspective Drawing
  Round 4: Verbal Description

Generalizing and Applying

- What are some examples of communications that you have seen others use?
- What are some careers that use drawings and graphics?
- What kind of things would you need to describe if you were a mechanical engineer, civil engineer, or an architect?
- What if you needed to draw something like an engine? What would be some ways of drawing and describing something that has internal parts?
- Youth can apply what they have learned in all design activities.

CAREER CONNECTIONS

Career Connection 1: Robotics Notebook

Scientists conduct experiments and try to find new knowledge. Engineers apply their knowledge of science, math, and other elements to solve problems. As they work, they need to record their ideas and progress. These notes are usually kept in a notebook. This notebook is an important tool for communication. It records what has been done, what has worked, what has not worked, and ideas about what to work on next. The notebook is also used to assign credit for discoveries, inventions, and patents.

- What type of experiments do you think scientists might work on and record in their notebooks?

- What might an engineer invent or design that would be recorded in a notebook? Engineers use both drawings and words to describe their ideas. Do you think it would be easier for you to use words or drawings to describe a new invention?

Activity C – Build Like a Technician

Performance Task For Youth

You will use the engineering design process to complete a building challenge that involves using manufacturing processes and design shapes.

Success Indicators

Youth are able to use the engineering design process in completing a challenge that includes using basic design shapes and tools.

List of Materials Needed

- Robotics Notebook
- Craft sticks with holes, brass paper brads, paper clips, sheet of standard printer paper (8 1/2 x 11)
- Basic tools: scissors to cut paper, punch or drill to make holes in craft sticks, pliers to grab or bend items like paper clips
- Optional: Poster/Handout of Manufacturing Processes on page 27
- Optional: Poster/Handout of Design Shapes on page 28
- Optional: Poster/Handout of the instructions on page 29

Activity Timeline and Getting Ready

- Activity will take approximately 20 minutes.
- Divide youth into groups of two or three.
- Try to divide parts so that each group gets a similar set of a variety of parts.

Safety Note

Before having the youth use the tools, discuss safety regarding their use. To learn more, visit: *www.4-H.org/curriculum/robotics.*

Experiencing

1. Lead a discussion on design and manufacturing elements. Use background information in the *Note to Leader* on page 13.

   Start with a discussion about how things are made. Some are natural; others are man-made; shaped; or assembled. Ask the participants to give examples of each.

   a. Example: Shelter – a cave is a natural shelter, a house is a man-made shelter.

   b. Use the posters or the Robotics Notebook section on manufacturing processes to discuss how man can shape or change natural items for our use, including building robots.

   c. Ask youth if they have used tools to do any of these processes. Have they used scissors, glue, a saw, or a drill? Show them some of the tools that you have available for their use. **Share or demonstrate the safe use of the tools.**

Building Round 1 - Drawing

2. Basic Square – Using a top, front, side view drawing (orthographic), have the participants construct the drawn item from the parts supplied. Allow them to use tools if needed.

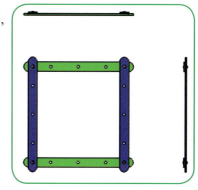

Building Round 2 - Drawing

3. Basic Triangle – Using a 3-D pictorial-type drawing (isometric), have the participants construct the drawn item from the parts supplied.

4. Lead discussion on how the shapes they made would work in building items. (The square may move into a parallelogram shape if the brads are not too tight. The triangle shape will be

very rigid and not move around.) When would you want something to move like the square and when would you want something rigid like the triangle? How could you make the square be rigid?

Ask the participants to think about other shapes they have seen in buildings, in cars, in planes, and in other items. What do they see besides triangles? Then share the poster or notebook page on design shapes such as cylinders, triangles, rectangles, arches, and others that are used in building things.

Building Round 3 - Description

5. Construct a support to hold a textbook above a tabletop using one sheet of printer paper and four paper clips. The book needs to be at least 5 inches above the tabletop. Use design elements and tools to create your support structure.

6. The youth should:

    a. Write the problem constraints – tasks and materials

    b. Draw many design ideas (at least three ideas)

    c. List what is good and bad about each idea

    d. Select and draw a final design

    e. Build it; try it out. Did it work?

    f. Write about modifications, shapes used, etc.

Engineering Design Process

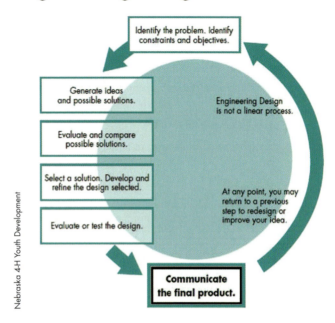

Nebraska 4-H Youth Development

Sharing and Processing

As the facilitator, help guide youth as they question, share, and compare their observations. Before they share with the group, have youth reflect on the activity in their Robotics Notebook. You may choose one of the questions below as a prompt. If necessary, use more targeted questions as prompts to get to particular points.

- What knowledge did you need in this building exercise?
- How did you know how to build the item?
- What determines the design shapes or manufacturing elements when constructing an item?

Generalizing and Applying

- Ask the youth to determine the design shapes of other items they have at home or see around the room.
- Ask youth to determine the manufacturing processes that were used in making some of their items around home, school, etc.
- Youth can apply what they have learned in Activity D.

Common Manufacturing Processes

Separate – Cut
Shear
Turn
Saw
Mill

Remove – Shape
Punch
Drill
Sand
Grind
Rout

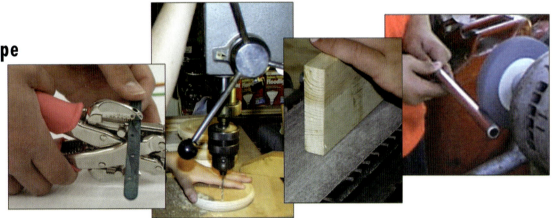

Bend – Form
Fold
Seam
Roll
Bend
Forge
Cast

Join – Fasten
Glue
Nail
Screw
Weld
Solder

Can you find the shapes?

Design Shapes

Triangle
Diamond
Rectangle - Square
Column - Tube
Arch - Circle

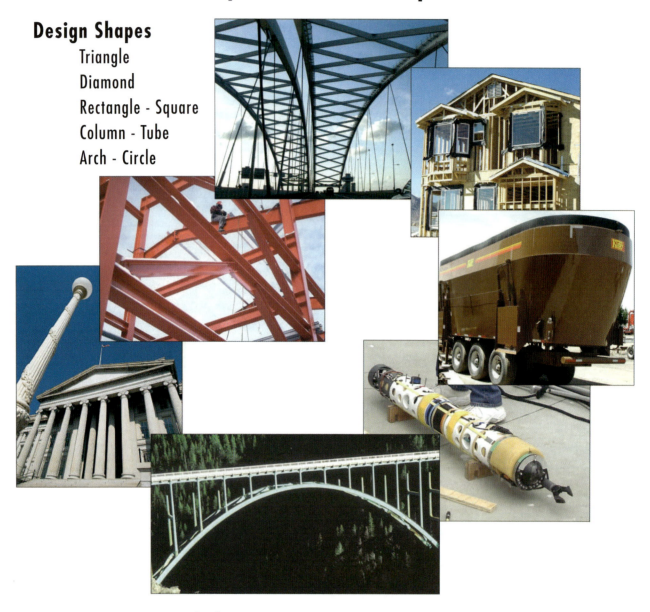

Material Shapes

L - Angle
I - Beam
Channel
Pipe
Round
Flat
Bar
Board
Sheet

Building Round 1 - Drawing – Basic Square – Using the top, front, side view drawing (orthographic drawing), construct the drawn item from the parts supplied.

Building Round 2 - Drawing – Basic Triangle – Using the 3-D picture-type drawing (isometric), construct the drawn item from the parts supplied.

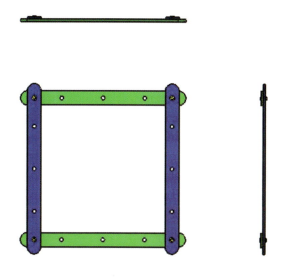

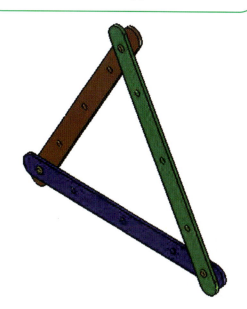

Building Round 3 - Description – Book Support – Construct a support to hold a textbook above a tabletop using one sheet of printer paper and four paper clips. The book needs to be at least 5 inches above the tabletop. Use design elements and tools to create your support structure.

Engineering Design Process

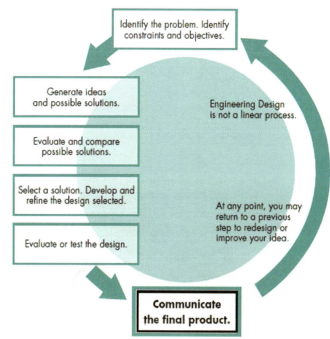

Identify the problem. Identify constraints and objectives.

Generate ideas and possible solutions.

Evaluate and compare possible solutions.

Select a solution. Develop and refine the design selected.

Evaluate or test the design.

Engineering Design is not a linear process.

At any point, you may return to a previous step to redesign or improve your idea.

Communicate the final product.

Use your own words and drawings to help you design and build the book support.

1. Write in your words the problem constraints – tasks and materials.

2. Draw many design ideas.

3. List what is good and bad about each design.

4. Select and draw a final design.

5. Build it; try it out. How did it work?

6. Write about modifications and shapes used in the final product.

Nebraska 4-H Youth Development

Activity D – Marshmallow Catapult Design Team

Performance Task For Youth

You will plan and design a swinging arm trebuchet-style catapult to launch marshmallows.

Success Indicators

Youth will be able to record thoughts, ideas, and design plans for a trebuchet-type catapult.

List of Materials Needed

- Robotics Notebook
- Collection of parts, for display only, of potential materials that can be used in creating the catapult design. These items include paper clips, binder clips, rubber bands, clothespins, craft sticks or paint paddles, paper brads, coffee stirrers (round, plastic tube style), drinking straws, and other items.
- Each team may need 7-10 paint-stirrer sticks and a small wooden dowel for their catapult.
- Optional materials include whiteboard, poster pad, or newsprint.
- Optional: Poster/Handout of Marshmallow Catapult Challenge on page 32

Activity Timeline and Getting Ready

- Activity will take approximately 20 minutes.
- Divide youth into teams of two or three.

Note to Presenter

The completed swing arm trebuchet catapults should be able to be used with slight modification as balance beam scales in the next module's activity, Sense of Balance.

Experiencing

1. Display/show materials to be used for this activity. Design Teams may only look at – but not touch or play with – items that they will be using during this design stage.
2. Review with the group the concepts of design and manufacturing elements and of form and function.
3. Give limited instructions on the design task. Design a catapult that satisfies the following requirements:
   - Design a trebuchet-style swing arm catapult.
   - The arm of the catapult must be adjustable (moveable pivot point).
   - Use at least five (5) different types of materials or parts (craft sticks, brads, dowels, etc.).
   - Use weights and gravity as the power source.
   - Launch a marshmallow at least six (6) feet.
4. Allow Design Teams to discuss how to make the marshmallow catapult.
5. Have youth use their Robotics Notebook – Activity D, Marshmallow Catapult Design Team to plan and draw their ideas.
6. Ask the Design Teams to share what they have designed with the entire group.

Sharing and Processing

As the facilitator, help guide youth as they question, share, and compare their observations. Before they share with the group, have youth reflect on the activity in their Robotics Notebook. You may choose one of the following questions as a prompt. If necessary, use more targeted questions as prompts to get to particular points. Remember, these questions are not about getting one right answer.

- What are some ways to assemble the parts?
- What types of shapes (construction methods) make strong but light structures?
- What other parts might make it easier to build this catapult?

Generalizing and Applying

- Encourage youth to consider other things that they could design or plan. What would they be?
- Youth can apply what they have designed in Activity E.

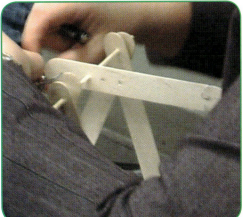

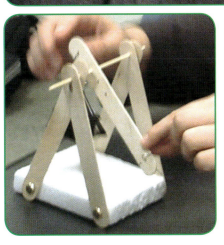
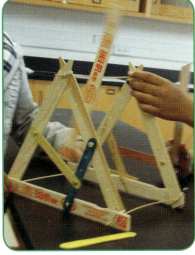
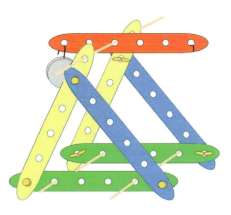

Marshmallow Catapult Challenge Design Task

Design requirements:

- Design a trebuchet-style swing arm catapult. It may be similar to a teeter-totter or swing set. The beam (arm) is able to swing (go up and down).

- The arm of the catapult must be adjustable from the center to an offset (One side of the beam is longer from center pivot point.).

- Use at least five (5) different types of materials or parts (craft sticks, brads, dowels, etc.).

- Use weights and gravity as the power source.

- Launch a marshmallow at least six (6) feet.

Activity E – Marshmallow Catapult Build Team

Performance Task For Youth

You will assemble parts, use simple tools, make modifications, and record information in your Robotics Notebook as you build a catapult.

Success Indicators

Youth will be able to use tools and parts to build a catapult of their design.

List of Materials Needed

- Robotics Notebook
- Collection of parts – Items that were on display in Activity D will be used to make the catapult.
- Optional materials include whiteboard, poster pad, or newsprint.
- Optional: Poster/Handout of Marshmallow Catapult Challenge on page 32

Activity Timeline and Getting Ready

- Activity will take approximately 30 minutes.
- Use the same groups from Activity D, Marshmallow Catapult Design Team.

Experiencing

1. Ask the Build Teams to build a catapult that will launch a marshmallow. Provide each group with the parts to be used in the construction. Use the ideas from the Design Team to build a catapult from the parts.

   a. Build Teams may do testing and modifications.

   b. When the catapults are complete, have Build Teams present and demonstrate how they were able to build and launch their catapult.

   c. They should record how close they came to launching the marshmallows 6 feet and measure five or more shots in a row for data.

   d. Build Teams should make notes in their Robotics Notebook, Activity E, to record actions and modifications (what worked, what didn't work, how it was modified).

Sharing and Processing

As the facilitator, help guide youth as they question, share, and compare their observations. Before they share with the group, have youth reflect on the activity in their Robotics Notebook. You may choose one of the following questions as a prompt. If necessary, use more targeted questions as prompts to get to particular points. Remember, these questions are not about getting one right answer.

- What design shapes worked well? (example: columns)
- What design shapes did not work as well?
- What are some other ways these catapults (trebuchets) could be made?
- What are some common classification groupings for catapults made in this exercise?
- Which method was easier for you to build the same catapult — written instructions or a diagram or both?

Generalizing and Applying

- Youth should try building other items to see how the shapes can support the items they build.
- Youth can apply what they have learned in the next module on arms.

Module 2: In Arm's Reach

Overview of Activities in this Module

To Learn
Activity F – Sense of Balance
Activity G – ABC ... XYZ

To Do
Activity H – Arm in Arm Design Team

To Make
Activity I – Arm in Arm Build Team

To Learn
Activity J – Pumped Up

To Do
Activity K – Just Add Air Design Team

To Make
Activity L – Just Add Air Build Team

Note to Leader

In this module, participants will explore different robot arm configurations and movements as well as make or assemble a robot arm. Following discussion of the different types of robot arms, members will suggest arm types by sketching and designing in the Robotics Notebook. Then, the participants will build mock-ups of their design. Use of levers and balance will be highlighted in the explorations and design of their robot arm. Participants will become familiar with locations in a three-dimensional space using X, Y, Z coordinates by playing a simple tic-tac-toe-style game.

The arm is one of the most useful and imitated parts of the human body. A robot arm can be used to grab, lift, move, or position items into a machine, to control a spot welder, or to assemble parts for an item.

The robot arm has six basic movements, three for the arm and three for the wrist.

Arm
- Vertical - up/down
- Radial - in/out
- Rotational - turning

Wrist
- Roll - swivel/rotate
- Pitch/Bend - up/down
- Yaw - side/side

There are four main ways to build a robot arm:

1. Polar coordinate – hinged arm that can go up and down, in and out, and pivot at the hinge; jointed arm hinged in two places, resembling a human shoulder and elbow.

2. Cylindrical coordinate – sliding arm that moves up and down on an upright tube, can rotate, and can go in and out, such as when slicing a pie.

3. Cartesian coordinate – straight movements using (X, Y, Z) coordinates for up and down, in and out, and back and forth movements in a cube or rectangular box shape.

While we identify and locate any point in 3-D space using the X, Y, Z coordinates, many robot arms use articulated or angled joints for their movement. In our four types of arm design, we can find both articulated angular movements and Cartesian linear movements. We can find three angular movements in the Jointed Arm; two angular and one linear movement in the Polar Arm; one angular and two linear movements in the Cylindrical Arm; and three linear movements in the Cartesian Arm. Articulated arm movement requires more programming and computation using the angles but provides other flexibility in movements.

What you will need for Module 2: In Arm's Reach

- Robotics Notebook for each youth
- Trunk of Junk, see page 8.
- Activity Supplies
  - Cardboard tubes, about one to two per group
  - Box of paper clips, at least one per group
  - Binder clips in different sizes, about 10-15 per group
  - Paper brads, about 5-10 per group
  - Rubber bands
  - Assortment of plastic squeeze bottles; water, ketchup, or basting bulbs/syringes for use in puffing air
  - Toy pinwheels or windmills, one per group
  - Toy balloons, two to three per group
  - Paper towels
  - Wooden sticks with pre-drilled holes
  - Paint sticks, about 5-10 per group
  - Craft sticks, about 10-15 per group
  - Cardboard or pegboard, around 6 inches by 12 inches, about one to two per group
  - Wood dowels, about two to four per group
  - Plastic syringes size 12cc to 30cc, six per group
  - Plastic tubing – ¼ inch (fish tank air tubing), 3-4 feet per group
  - Metal bolt washers (used as weights), 1-inch outside diameter works well, about 5-10 per group
  - Optional – hinges, about two to three per group
  - Ruler to measure inches
- Tool Box
  - Glue
  - Tape
  - Scissors, one or two per group
  - Low-temperature glue gun
  - Hand drill with bits
  - Small hacksaw (to cut dowels and small boards)
- Things to make or acquire
  - 3-D Tic-Tac-Toe game board (purchase or make your own), one per two teams
  - 3-D Tic-Tac-Toe spinner (instructions to make on page 45), one per two teams
  - 3-D Tic-Tac-Toe game markers (purchase or make your own), six or more per team
  - Optional: Copies of the scorecard on page 47 for each team

Timeline for Module 2: In Arm's Reach

Activity F – Sense of Balance
- Activity will take approximately 50 minutes.
- Divide youth into small groups of two or three.
- Use marshmallow catapult from Activity E for each group.
- Set out enough washers and paper clips for each group to test balance with various pivot points.

Activity G – ABC ... XYZ
- Activity will take approximately 20 minutes.
- Divide youth into small teams of two to three. Two teams will play together on one 3-D board.

Activity H – Arm in Arm Design Team
- Activity will take approximately 20 minutes.
- Divide youth into small groups of two or three.
- Materials for this activity are for display only.

Activity I – Arm in Arm Build Team
- Activity will take approximately 40 minutes.
- Use same grouping from Activity H – Arm in Arm Design Team.

Activity J – Pumped Up
- Activity will take approximately 30 minutes.
- Divide youth into small groups of three or four.

Activity K – Just Add Air Design Team
- Activity will take approximately 20 minutes.
- Divide youth into small groups of three or four.

Activity L – Just Add Air Build Team
- Activity will take approximately 40 minutes.
- Use same grouping from Activity K – Just Add Air Design Team.

Big Ideas

- Use of simple machines
- Form and function
- Three-dimensional space is critical in robot arm movement.

NSE Standards

- Systems, order, and organization
- Abilities of technological design

STL

- Core concepts of technology
- Relationships and connections
- Attributes of design
- Manufacturing technologies

Performance Tasks For Youth

You will experiment in balancing unequal weights on a balance beam while also moving the pivot point to different locations on the beam.

This activity will help you understand the three axes of a cube, X, Y, Z, and locations in 3-D space.

You will design and draw a robot arm that you will build, using levers to pick up and move a weight from one spot to another location. The arm will have at least two of the three axes of movement, X, Y, Z.

You will build a robot arm from your design in Activity H.

You will explore moving objects with balloons, plastic bottles, and syringes.

You will design a power source to move the arm you built in Activity I.

You will use your plans from Activity K and add a power source to move the arm built in Activity I.

SET Abilities

- Compare/Contrast
- Communicate/Demonstrate
- Draw/Design
- Build/Construct

Life Skills

- Keeping Records
- Planning/Organizing
- Critical Thinking
- Problem Solving
- Communication
- Contribution to Group Effort
- Teamwork

Success Indicators

Youth are able to demonstrate various methods to balance a beam using washers and paper clips.

Youth are able to place markers on the correct location and level in a 3-D space. Youth are able to determine and understand when a straight line of three is accomplished in 3-D.

Youth will understand the principles behind arms and movement and use them as a base for the designs they draw for a robotic arm that they will build.

Youth will understand the principles behind arms and movement as demonstrated by building a robotic arm using levers to pick up and move a weight from one spot to another location.

Youth can use air power (pneumatics) as a focus power source to lift and move simple objects.

Youth will understand the use of pneumatics as power sources by being able to design and sketch a plan to attach an air power source to move their robotic arm.

Youth will have applied the use of pneumatics by building and attaching a system to move the robot arm they built in Activity K.

Activity F – Sense of Balance

Performance Task For Youth

You will experiment in balancing unequal weights on a balance beam while also moving the pivot point to different locations on the beam.

Success Indicators

Youth are able to demonstrate various methods to balance a beam using washers and paper clips.

List of Materials Needed

- Robotics Notebook
- Marshmallow catapult (one for each group of youth) built in Activity E or parts and items to be used as levers and fulcrums (pivot point) to act as a balance scale
- Metal bolt washers as weights; 1-inch outside diameter washers are good sizes, but any kind or size will work. About 5-10 washers per group
- Four to six smooth paper clips per group
- Optional: Copies of data sheet on page 40 for each group
- Ruler to measure inches

Activity Timeline and Getting Ready

- Activity will take approximately 50 minutes.
- Divide youth into small groups of two or three.
- Modify catapults used in Activity E so they can be used as balance beams.
- Set out enough washers and paper clips for each group to test balance with various pivot points.

You should be able to modify the Marshmallow Catapult and use it as a balance scale. Have a discussion about levers and leverage as part of this activity. In Activities M, N, and O, there will be additional discussion on types of levers related to grippers.

Experiencing

1. Have one balance beam for each team. Have participants move the pivot point by removing the pivot (dowel) and placing it in a hole near the end of the stick (beam) while leaving at least one hole on either side of the pivot.

2. Ask teams to perform the following actions with their beam:

   a. Balance the beam horizontally by only placing the washers and paper clips on one hole on either side of the pivot. Hang the washers on a bent paper clip "S" hook. Record in the Robotics Notebook the number of washers (weight) used and the number of inches (distance) from the pivot to balance the beam. If it can't be balanced, record the locations and number of washers at the position closest to being balanced.

   b. If the beam could not be balanced, add paper clips to the hooks with the washers to provide fine adjustment to balance the beam.

   c. Move the pivot to another location on the beam. Again using the metal washer weights and paper clips, balance the beam at this new location. Record in the Robotics Notebook the number of washers (weight) and the number of inches (distance) from the pivot used to balance the beam. Repeat for each change of pivot point.

3. Have teams share findings; use poster paper and have teams chart the results of balance on the beams.

4. Challenge: Use the data collected in the Robotics Notebook — distance and weight — and multiply the number of washers by the distance the washers are located. This will give you the amount of torque that is present on either side of the beam. When the torque on one side of the beam is equal to the amount of torque on the opposing side, the beam will be balanced. Use the torque equation below to find the torque of each side of the beam and compare the results among the teams.

Torque equations:

Torque	= Force X	Distance
(number of washers)	X	(inches from the pivot)

Balanced Beam:
Torque on left side = Torque on right side

Example:
(3 washers) **X** (2 inches) **=** (2 washers) **X** (3 inches)

$$3 \times 2 = 2 \times 3$$
$$6 = 6$$

Note: the weight calculations on either side of the pivot point may not equal due to limited accuracy in measurements and exact weights of the washers.

5. Have participants record in their Robotics Notebook their experiences, their discoveries, and what they learned.

As the facilitator, help guide youth as they question, share, and compare their observations. Before they share with the group, have youth use their Robotics Notebook to record ideas, comments, and notes on the activities they have been doing. You may choose one of the questions below as a prompt. If necessary, use more targeted questions as prompts to get to particular points.

- What are some ways to make a balance beam?
- When you change the pivot point on the balance beam, what happens?
- Describe some common levers we use (e.g., wheelbarrow, crowbar, bottle opener).
- What levers are similar to the one you made?
- How are the other levers different?

Generalizing and Applying

- Try balancing on a teeter-totter. Can you balance it with two groups of people of unequal weight on the sides? If so how?
- How can you use this information to build a robot arm (lever)?
- Why is it important to consider torque?

Data Sheet

Use graphic datasheet for handout if not using the Robotics Notebook.

	Sense of Balance Data Sheet						
	Left Side			Balanced	Right Side		
	Weight (Number of Washers)	**Distance** (Pivot to Washers) (Inches)	**Torque** (Weight x Distance)	**=**	**Weight** (Number of Washers)	**Distance** (Pivot to Washers) (Inches)	**Torque** (Weight x Distance)
Example	4	3	12	=	6	2	12
1				=			
2				=			
3				=			
4				=			
5				=			
6				=			
7				=			
8				=			

Data Sheet

Use graphic datasheet for handout if not using the Robotics Notebook.

	Sense of Balance Data Sheet						
	Left Side			Balanced	Right Side		
	Weight (Number of Washers)	**Distance** (Pivot to Washers) (Inches)	**Torque** (Weight x Distance)	**=**	**Weight** (Number of Washers)	**Distance** (Pivot to Washers) (Inches)	**Torque** (Weight x Distance)
Example	4	3	12	=	6	2	12
1				=			
2				=			
3				=			
4				=			
5				=			
6				=			
7				=			
8				=			

Activity G – ABC ... XYZ

Performance Task For Youth

This activity will help you understand the three axes of a cube, X, Y, Z, and locations in 3-D space.

Success Indicators

Youth are able to place markers on the correct location and level in a 3-D space. Youth are able to determine and understand when a straight line of three is accomplished in 3-D.

List of Materials Needed

- Robotics Notebook
- One 3-D game board, spinner, and game markers for two teams of youth to play against each other
- Copy of 3-D Tic-Tac-Toe Game Instructions on page 48 for each group
- Optional: Copies of the scorecard on page 47 for each group

Activity Timeline and Getting Ready

- Activity will take approximately 20 minutes.
- Divide youth into small teams of two to three. Two teams will play together on one 3-D board.
- Make a spinner for each game board using the template and instructions on page 45.
- Purchase the 3-D game board or create your own using recycled CD cases and the template on page 46. See page 44 and the 4-H Robotics website for more details and suggestions at *www.4-H.org/curriculum/robotics*.
- If you make your own game board, you also will need to gather game markers. See page 44 for details.

Experiencing

1. Have teams play the 3-D tic-tac-toe game. Each team spins the spinner to get each of the three coordinates (X, Y, Z) to place their marker. At each spin, members record the X, Y, Z values in their Robotics Notebook. When a team can make a straight line with their markers on any level, they win.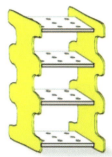

2. Try rounds with youth playing on different teams.

3. If using teams of three, one player spins, one member records spinner coordinates, and the third player places the marker on the location. All team members confirm the 3-D location on the game board.

Sharing and Processing

As the facilitator, help guide youth as they question, share, and compare their observations. Before they share with the group, have youth use their Robotics Notebook to record ideas, comments, and notes on the activities they have been doing. You may choose one of the questions below as a prompt. If necessary, use more targeted questions as prompts to get to particular points.

- Have teams discuss the different ways they could make a straight line.
- How many points would it take to make a straight line?
- What would happen if there were more points on the grid? (accuracy, complexity)

Generalizing and Applying

- What are some other things that use or have three dimensions?
- Youth can be asked to apply the 3-D sorting of things in a cupboard or other 3-D area.
- Why is thinking or seeing in 3-D important to us? To a robot?
- Youth also can apply what they have learned in Activity H.

CAREER CONNECTIONS

Career Connection 2: Attributes of an Engineer

Being an engineer requires specific knowledge and skills, above and beyond just training in engineering fields. While engineers must have a good understanding of basic scientific knowledge, including mathematics, physical and life sciences, and information technology, they also need to be curious, creative, and have a desire to learn.

Engineers need to be well-rounded and understand the social context of their issue, including the history, economics, and environment relating to the problem. Basic knowledge of science and society provides engineers with a foundation for approaching potential solutions that will work best in the real world.

In addition, engineers often have many natural characteristics that prepare them well for their career. Engineers need to understand the importance of teamwork and the ability to work with others. Engineers must be patient and flexible. They should have sound ethics, good communication skills (written, verbal, and graphics), and have safe work practices. These traits, combined with their formal education, help engineers to work effectively when devising innovative solutions.

- Which of these life skills do you possess? Which do you need to improve?
- In your opinion, what is the most important skill for an engineer? Why?

3-D tic-tac-toe

This game is designed to help youth understand the three axes of a cube —
X, Y, Z — and locations in 3-D space. Robot arms must be able to move and
locate positions in a 3-D space.

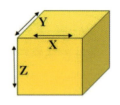

This game board can be purchased on-line or you can make your own game
board using pieces of clear plastic. Used containers, CD cases, or other items
that you can find can work if you make your own. The graphics included in this
guide should work with most CD cases.

Parts needed:

Game Stand

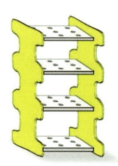

Purchase a four level game board on-line and label each level (Z) with X,
Y grids, or make your own game board. See the web site at *www.4-H.org/
curriculum/robotics* for hints on making your own 3-D game boards.

Game Wheel and Spinner

(The wheel and spinner must be made as they do not come with purchased
stand.)

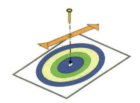

The Game Wheel is designed to provide the three (X, Y, Z) coordinate numbers
with each spin. There are three rings with numbers (X, Y, Z). Different
combinations of X, Y, Z can then be read from the spinner and marked on the
scorecard. Marking on the scorecard ensures that it was read correctly and can
be used to verify proper placement of each team's markers.

The Game Spinner can be cut from cardboard. A popsicle stick with a hole also
could be used. The spinner should turn freely (spin) and stop at different places
on the wheel.

Team Markers

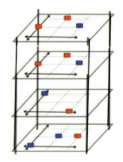

Small items that do not move around and that can stack make good markers.
Buttons might be good, while marbles would not work well. Each team should
have different colors, shapes, or some way to distinguish which markers belong
to each team.

Game Rules and Scorecards

Make copies of the scorecard and rules on pages 47 – 48, so that each team has
a copy or have youth use the Robotics Notebook to keep coordinates for each
team spin.

Game Wheel

Photocopy this page with the game wheel template and attach it to a backing material. Make a hole in the center for the spinner.

- Game wheel is to be printed or copied (in color, if possible, for easier reading of X, Y, Z coordinates).
- The wheel is formatted to fit the bottom of a CD case if printed about 4 inches in diameter, but can be attached to any stiff material (plastic, cardboard, etc.).
- Adjust copy size, if needed, if using different material for backing.

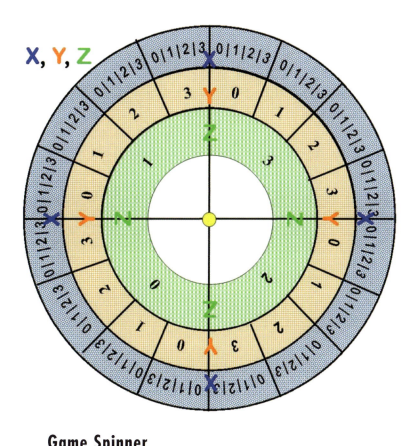

Game Spinner

Make a spinner for the game wheel from used cardboard using the pattern below or use a drilled craft stick as a spinner.

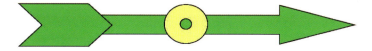

X,Y Grid - template

Use a prebuilt game board or make a game using old CD cases. Then copy or print out the template below on paper and trace the grid onto the top clear plastic levels of the game board. Attach or mark the printed X,Y grid directly onto the bottom level.

The grid is formatted to fit a used CD case if printed as a 4 inch by 4 inch grid.

If using a premade game, adjust the template size to fit the game board size.

For the Z coordinate, mark each level as appropriate, 0 is the base level and 1, 2, and 3 are the upper levels.

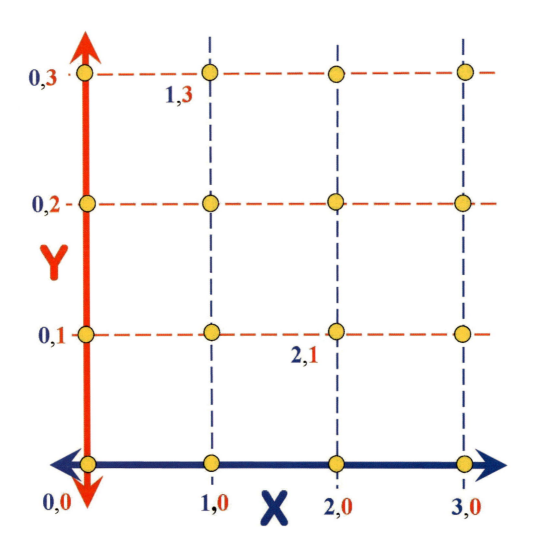

3-D Tic-Tac-Toe Game Scorecard

Team Name: _____

Turn	X	Y	Z	
Spin 1				
Spin 2				
Spin 3				
Spin 4				
Spin 5				
Spin 6				
Spin 7				
Spin 7				
Spin 8				
Spin 9				
Spin 10				
Spin 11				
Spin 12				
Spin 13				
Spin 14				
Spin 15				

Team Name: _____

Turn	X	Y	Z	
Spin 1				
Spin 2				
Spin 3				
Spin 4				
Spin 5				
Spin 6				
Spin 7				
Spin 7				
Spin 8				
Spin 9				
Spin 10				
Spin 11				
Spin 12				
Spin 13				
Spin 14				
Spin 15				

3-D Tic-Tac-Toe Game Instructions

1. Divide the group into teams of two to three players.

2. Have two to four teams per 3-D game board.
   - Have one member from each team spin the spinner to determine the order of play. The team with the highest number starts first. If there is a tie, teams respin until the tie is broken.
   - Play begins by having each team spin to get the three coordinate numbers (X, Y, Z).

3. Each team records the X, Y, Z numbers on its scorecard for that spin.

4. The team then finds the 3-D location on the 3-D stand and places one of its game pieces on that spot.

5. Play continues with each team spinning and marking its 3-D location.

6. If a game piece marker is already on that spot:
   - the team may place its marker on top of the other piece, provided the piece on the spot belongs to another team;
   - if the team's own marker is on the spot, the team should spin again until it gets an empty spot.

7. The goal is to get at least three markers to align in a straight row.*

8. The three-in-a-row can be on the same level, up and down, or at an angle.

   * The question of what determines a straight line can be discussed by the group to determine if the markers are in a line.

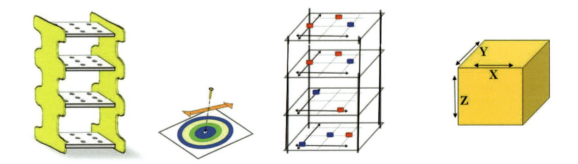

Activity H – Arm in Arm Design Team

Performance Task For Youth

You will design and draw a robot arm that you will build using levers to pick up and move a weight from one spot to another location. The arm will have at least two of the three axes of movement, X, Y, Z.

Success Indicators

Youth will understand the principles behind arms and movement and use them as a base for the designs they draw for a robotic arm that they will build.

List of Materials Needed

- Robotics Notebook
- Collection of parts, for display only, of the potential materials that can be used in creating the robot arm design. Items that may be part of the Trunk of Junk collection include paper clips, binder clips, clothespins, craft sticks, paper brads, coffee stirrers (round, plastic tube style), and drinking straws. Also, items such as paint sticks and cardboard tubes from aluminum foil and paper towels can be used.
- Optional: Whiteboard, poster pad, newsprint, etc.
- Optional: Copies of handouts/posters on robot arm movement and designs on pages 51-52.

Activity Timeline and Getting Ready

- Activity will take approximately 20 minutes.
- Divide youth into small groups of two or three.
- Materials for this activity are for display only.

Experiencing

1. Lead the group in discussion.

   a. What is an arm? Ask for a definition of an arm. (Examples: human arm, arm to set the car alarm, arm on a chair, arm used to mean to carry a rifle.) What can you do with an arm? How does a monkey's or kangaroo's arm or the crossing arm at a railroad track differ? What do they have in common?

   b. Using a blackboard or poster paper, have youth draw shapes of the different arms mentioned in the discussion.

   c. What is a robot arm? Ask the group for examples of robot arms or things that look like arms. Examples may include an automated car wash, tractor backhoe, garbage truck bin dumping arm, and forklift. Do these have the same purpose or use as the arms listed in question 1?

   d. The robot arm has six basic movements: three for the arm and three for the wrist.

   Arm
   - up/down
   - in/out
   - turning

   Wrist
   - Roll – roll/rotate
   - Pitch – up/down
   - Yaw – side/side

   How do robots' arms move?

e. There are four main ways to construct a robot arm:

    i. Two-hinged arm – most like our arms; jointed arm

    ii. One-hinged arm – polar coordinate

    iii. Sliding arm on an upright tube – cylindrical coordinate

    iv. Straight movement arm – Cartesian coordinate

2. Divide into design groups.

a. Have the Design Teams use the Activity H page in the Robotics Notebook to sketch their robot arm designs.

b. Using the parts available, have the Design Teams design and draw a robot arm they can build that uses levers to pick up and move a weight from one spot to another location.

c. This will be a two-part challenge. For this meeting, the challenge will be to design an arm that will move in at least two of the three coordinate directions: 1. side to side (X), 2. in/out (Y), and 3. up/down (Z), and will lift and move a weight to a different location.

d. The second part of the challenge will be in Activities K and L in which youth will learn about pneumatics and how to add air power to move the robot arm they designed.

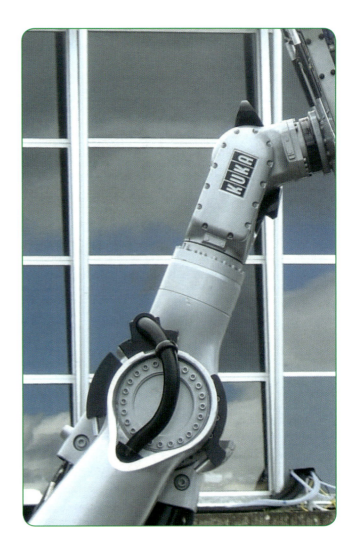

Sharing and Processing

As the facilitator, help guide youth as they question, share, and compare their observations. Before they share with the group, have youth use their Robotics Notebook to record ideas, comments, and notes on the activities they have been doing. You may choose one of the questions below as a prompt. If necessary, use more targeted questions as prompts to get to particular points.

- How many degrees of freedom (direction of movement) did your arm have?

- What type of arm did you make: cylindrical coordinate, polar coordinate, Cartesian coordinate, or jointed?

- What type of arm would be best to pick up a heavy weight?

- How can you use levers to help pick up the weight?

- What other parts might make it easier to make this robot arm?

Generalizing and Applying

- Encourage youth to consider other things that they could design or plan that could use the X, Y, Z coordinates or the arm movement designs. What could they be?

- What methods were used to make the movements in other machines? Turning (rotating)? Sliding (linear)? Lever-like (articulating)?

- Youth also can apply what they have designed in Activity I.

Movements of the Robot Arm and Wrist

Arm

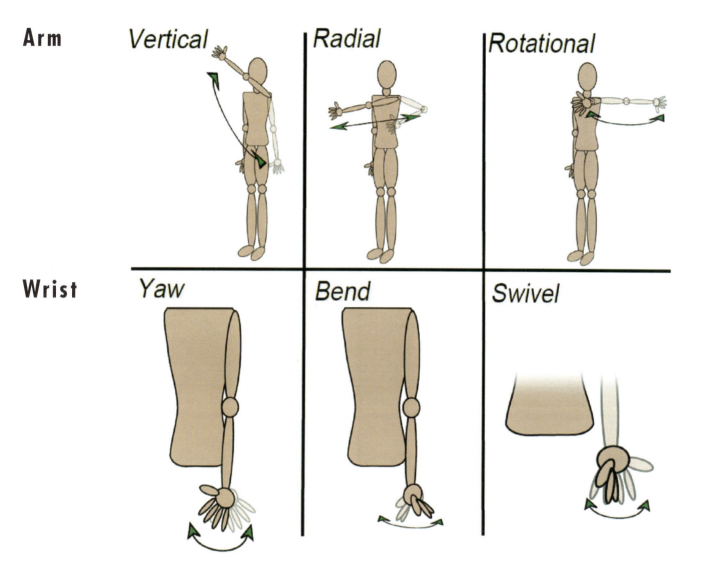

Vertical   Radial   Rotational

Wrist

Yaw   Bend   Swivel

Four Main Types of Robot Arm Designs

Note: The green shapes show the work area of the space that the arm is able to move within.

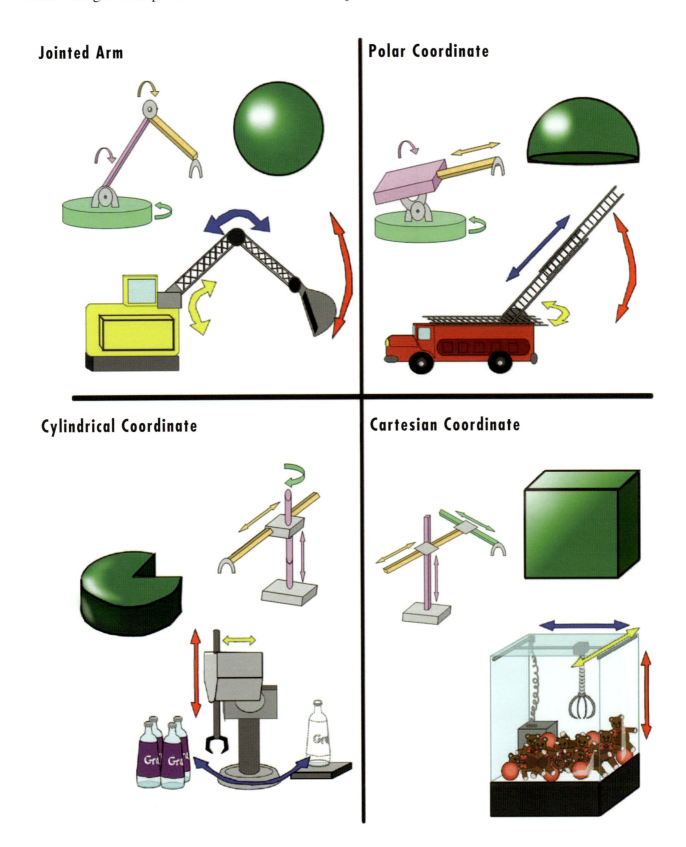

Jointed Arm

Polar Coordinate

Cylindrical Coordinate

Cartesian Coordinate

Activity I – Arm in Arm Build Team

Performance Task For Youth

You will build a robot arm from your design in Activity H.

Success Indicators

Youth will understand the principles behind arms and movement as demonstrated by their building a robotic arm using levers to pick up and move a weight from one spot to another location.

List of Materials Needed

- Robotics Notebook
- Collection of parts – items from the Trunk of Junk: drilled paint sticks, small trim boards, pegboard, straws, wood dowels, small bolts, paper brads, rubber bands, paper clips
- Other items: scrap parts, glue, bolts, screws, and other miscellaneous hardware or building kits, tape, structure parts, weights, and other supplies used before. (Making the arms from items the size of paint sticks, rulers, or yardsticks will make it easier to attach the syringes and grippers to be made and/or added to the arm in Activities Q and S.)
- Tables, cardboard bases, or pegboard building stands to support the arms as they are constructed
- Tools such as a low-temperature glue gun, hacksaw, hand drill with bits, scissors

Activity Timeline and Getting Ready

- Activity will take approximately 40 minutes.
- Use same grouping from Activity H, Arm in Arm Design Team.

Experiencing

1. Set out the materials. Have teams build the robot arms that they designed. Each robot arm must:

   a. Use levers.

   b. Be able to pick up a weight.

   c. Be able to move in **two** of the three coordinate directions:

      i. X – side to side
      ii. Y – in and out
      iii. Z – up and down

2. After building the robot arms, have the groups share the arms they built.

3. Build Teams should record actions and modifications in their Robotics Notebook.

Sharing and Processing

As the facilitator, help guide youth as they question, share, and compare their observations. Before they share with the group, have youth use their Robotics Notebook to record ideas, comments, and notes on the activities they have been doing. You may choose one of the questions below as a prompt. If necessary, use more targeted questions as prompts to get to particular points.

Ask the participants to describe how they made their arm:

- What type of parts did you use (fasteners, levers, etc.)?
- How do the arm movements of a robot act like our own arms?
- How are the arm movements of a robot different from our arms?
- What problems did you have building the arm? What worked well?
- Describe what you observed as you built the robot arm for this activity.
- What are the different types of robot arms and how do they work (move)?

Generalizing and Applying

- Identify the places we use robots. Are there any robotic arms in your house? How would your life be different if we didn't have robots? Why?

Activity J – Pumped Up

Performance Task For Youth

You will explore the moving of objects with balloons, plastic bottles, and syringes.

Success Indicators

Youth can use air power (pneumatics) as a focus power source to lift and move simple objects.

List of Materials Needed

- Robotics Notebook
- A selection of balloons, plastic soda/water bottles, kitchen bulb basters, syringes, and plastic tubing so that each group has some source of air power

- Each group should have washers for weights, two to four balloons, and one or two small pinwheels or other objects to move or twirl.
- Other items such as paper clips, paper brads, and thumbtacks from the Trunk of Junk also may be used.

Activity Timeline and Getting Ready

- Activity will take approximately 30 minutes.
- Divide youth into small groups of three or four.

Experiencing

1. Set items out on a table: balloons, plastic bottles with narrow necks, bottles with pop-up and open/close tops, pinwheels, and washer weights.
2. Give each group a bottle and a pinwheel.
3. Ask participants to use the bottle to blow on the pinwheel to make it spin.
   a. Record results: Which bottles worked well? Which didn't? How well did the wheels spin?
   b. How is this example like the wind? How easy is it to control this type of power?
   c. This is an open system, like that of a windmill used for pumping water or generating electricity.
4. Ask the teams to use air power to raise or move the weights or other objects, not by blowing on the items to move them, but by attaching a balloon to the bottle.
   a. Have teams try to raise the most weight or move the most distance using only air power from a bottle and a balloon.
   b. Have teams show and share how they accomplished the task.
   c. How did the balloon function in this activity? How easy is it to control this type of power?
   d. This is a closed system, such as one in an air compressor with tubing, valves, and cylinders.
5. Have participants record their experiences and what they learned in their Robotics Notebook.

Sharing and Processing

As the facilitator, help guide youth as they question, share, and compare their observations. Before they share with the group, have youth use their Robotics Notebook to record ideas, comments, and notes on the activities they have been doing. You may choose one of the questions below as a prompt. If necessary, use more targeted questions as prompts to get to particular points.

- What are some ways pneumatics (using controlled air) can be used to lift items?
- Describe some pneumatic devices used to do work in your community. Are they open or closed systems?
- Explain which of the pneumatic power applications used by the teams worked best.

Generalizing and Applying

- Where is — or how could — air power be used in a dairy, a grocery store, or other business?
- What if we used water in the tubing instead of air? How would that be different?
- Youth can apply what they have learned in Activity K.

Activity K – Just Add Air Design Team

Performance Task For Youth

You will design a power source to move the arm you built in Activity I.

Success Indicators

Youth understand the use of pneumatics as power sources by being able to design and sketch a plan to attach an air power source to move their robotic arm.

List of Materials Needed

- Robotics Notebook
- Collection of parts, for display only, of the potential materials that can be used in creating the air power design. Items can be part of the Trunk of Junk collection, such as paper clips, binder clips, clothespins, craft sticks, paper brads, and drinking straws. Items such as plastic syringes, plastic tubing, and connectors also can be used.
- Optional whiteboard, poster pad, newsprint, etc.

Activity Timeline and Getting Ready

- Activity will take approximately 20 minutes.
- Divide youth into small teams of three or four.

Experiencing

1. Lead the group in discussion.

   a. How do robots' arms move? The robot arm has three basic power sources:

   Electricity – very accurate, repeatable, but slower

   Hydraulics – very strong, incompressible, but messy

   Pneumatics – very quick, lightweight, but lower accuracy

   b. How these power sources are used in a robot arm:

   i. Electric - motors, gears, screws, pulleys, electromagnetic actuators
   ii. Hydraulics – use of liquids to perform mechanical tasks, pressurized fluid in tubing and cylinders
   iii. Pneumatics – use of gases to perform mechanical tasks, compressed air channeled through tubing and cylinders

2. Divide into design groups.

    a. Use plastic syringes and tubing for the air power to move the robot arm and weight from one spot to another location.

    b. Using the parts shown for display only, the Design Teams should design and draw a power system for a robot arm.

    c. Each robot arm must:

        i. Use air power and be able to accomplish the task without direct hand movement by the team.

        ii. Be able to lift a weight.

        iii. Be able to move in two of the three coordinate directions.

            1. X – side to side

            2. Y – in and out

            3. Z – up and down

3. Have the Design Teams use the Robotics Notebook to sketch their robot arm power designs.

Sharing and Processing

As the facilitator, help guide youth as they question, share, and compare their observations. Before they share with the group, have youth reflect on the activity in their Robotics Notebook. You may choose one of the questions below as a prompt. If necessary, use more targeted questions as prompts to get to particular points.

- What type of air power system did you design? (open system, closed system, multiple cylinders [syringes], etc.)
- Where did you position the cylinders on the lever arms?
- What are the different ways you can connect power cylinders (syringes) to move the arm?
- What other parts might make it easier to move this robot arm?

Generalizing and Applying

- Encourage youth to consider other things that they could design or plan that could use air power. What would they be?
- How would you change the design if you used water (hydraulics) instead of air (pneumatics) in your system?
- Youth also can apply what they have designed in Activity L.

Activity L – Just Add Air Build Team

Performance Task For Youth

You will use your plans from Activity K and add a power source to move the arm built in Activity I.

Success Indicators

Youth will have applied the use of pneumatics by building and attaching a system to move the robot arm they built in Activity K.

List of Materials Needed

- Robotics Notebook
- Collection of potential materials that can be used in creating the air power design; items can be part of the Trunk of Junk, such as paper clips, binder clips, clothespins, craft sticks, paper brads, and drinking straws.
- Activity supplies – plastic syringes, plastic tubing, and connectors also must be used.
- Tools and fasteners: low-temperature glue gun, hacksaw, scissors, bolts, screws, and other miscellaneous hardware or building kits, tape, structure parts, drilled craft sticks, pegboard, weights, and other supplies used in Activity J.

Activity Timeline and Getting Ready

- Activity will take approximately 40 minutes.
- Use the same groups from Activity K, Just Add Air Design Team.

Safety Note

Syringes hooked together with tubing can act like a pop gun, shooting a plunger from a syringe. Tell members not to do that!

Experiencing

1. Have teams build the power system and add it to the robot arms that they built in Activity I. Each robot arm must fulfill the following:

   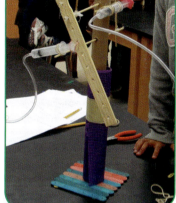

   a. Use air power and be able to accomplish the task without direct hand movement by the team.

   b. Be able to lift a weight.

   c. Be able to move in two of the three coordinate directions:
      i. X – side to side
      ii. Y – in and out
      iii. Z – up and down.

2. After building the robot arm with air power, have the groups share the arm they built.

3. Have youth record their observations in their Robotics Notebook.

Sharing and Processing

As the facilitator, help guide youth as they question, share, and compare their observations. Before they share with the group, have youth use their Robotics Notebook to record ideas, comments, and notes on the activities they have been doing. You may choose one of the questions below as a prompt. If necessary, use more targeted questions as prompts to get to particular points.

- How did you add air power to the arm?
- How do the arm movements work with air power? (smooth, jerky, varied)
- What problems did you have in adding air power to the arm? What worked well?
- How would you use a different power source? (electricity, hydraulics)
- Would a different power source be better for your arm?

Generalizing and Applying

- Try making an "air muscle" to pull — rather than push — with air. Use braided rope or a net bag (Chinese finger trap) over a balloon. Tie it at the ends but leave an opening for air flow in one end of the balloon.
- Try using a different power source to move the arm. An electrical motor turning a screw can move in a straight line. Try using a lip balm stick, glue stick, or container of deodorant to show linear movement on an arm.
- Youth also can apply what they have learned in the next module on robot hands and grippers.

Module 3: Get a Grip

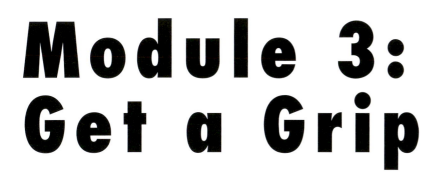

Overview of Activities in this Module

To Learn
Activity M – Chopsticks
Activity N – Just a Pinch
Activity O – Hold On

To Do
Activity P – One for the Gripper Design Team

To Make
Activity Q – One for the Gripper Build Team

To Do
Activity R – Twist of the Wrist Design Team

To Make
Activity S – Twist of the Wrist Build Team

Note to Leader

The robot hand, which is attached to the robot's wrist, also is called an "end effector." Some common uses of the hand include grippers to hold or pick up items. Grippers can be mechanical, suction cups, electromagnets, hooks, scoops, or ladles.

Other types of end effectors are tools that can be attached to the wrist or held by a gripper. Some common tools are welders, spray paint guns, drilling spindles, routers, grinders, wire brushes, and heating torches.

The motion of the robot arm can affect the type of hand that can be used on the arm. Two common types of motion for arms and hands are Point to Point (PTP) and Continuous Path. In Point to Point motion, the robot knows the starting and stopping points and will move in a direct path from one to the other. The PTP motion is good for loading or moving items, picking up and placing packages or parts, spot welding to each point or spot to make a weld, and similar processes.

In the Continuous Path motion, the robot follows a long set of points, usually right next to each other. Continuous Path motion is good for moving near a contour, such as paint spraying or bead welding. Continuous Path motion takes more programming code and directions from the computer.

Most mechanical grippers use some sort of joints and linkages to form the gripping action. Assembling may involve many parts, depending on the needs of the gripper. For example, grippers may need to carefully hold and carry an egg, or pick up a heavy metal part and position it for assembly on large equipment, or other tasks. Each type of gripper will need specific criteria and construction to accomplish the task it has been designed to perform.

What you will need for Module 3: Get a Grip

- Robotics Notebook for each youth
- Trunk of Junk, see page 8.
- Activity Supplies
  - Drinking straws, two for each participant
  - Chopsticks or small dowels, two for each participant
  - An assortment of pliers, tongs, scissors, and nutcrackers, enough for each participant or pair to have one
  - A selection of small items to pick up: marbles, balls, ping-pong balls, golf balls, plastic eggs, wood blocks; also include some items that cannot be picked up by the grippers you have available. Total number should be at least the same as the number of participants.
  - Various items and parts for a gripper, such as hinges, bolts, nuts, wood, plastic pieces, craft sticks, syringes, tubing, etc., from the Trunk of Junk
- Things to make or acquire
  - The robot arms the teams built in Module 2: Arm In Arm, Activities I and L
  - The gripper to be built in Activity Q
- Tool Box
  - Simple tools – saws, drills, and other tools

Timeline for Module 3: Get a Grip

Activity M - Chopsticks
- Activity will take approximately 15 minutes.
- Divide youth into small groups of two or three.
- Set parts on the table.

Activity N – Just a Pinch
- Activity will take approximately 20 minutes.
- Arrange youth around a table or circle of chairs. If it is a large group, you may set up multiple tables or circles.
- Collect enough grippers (tongs, pliers, etc.) so that each individual will have one.

Activity O – Hold On
- Activity will take approximately 20 minutes.
- Divide youth into small groups of two or three.

Activity P – One for the Gripper Design Team
- Activity will take approximately 20 minutes.
- Divide youth into small Design Teams of two or three.
- Display building items for viewing.

Activity Q - One for the Gripper Build Team
- Activity will take approximately 50 minutes.
- Use the same teams from Activity P – One for the Gripper Design Team.
- Make parts and supplies available for building.
- Provide a work area for tools and assembly.
- Activity P – One for the Gripper Design Team must be completed before doing this activity.

Activity R – Twist of the Wrist Design Team
- Activity will take approximately 20 minutes.
- Divide youth into small Design Teams of two or three.

Activity S – Twist of the Wrist Build Team
- Activity will take approximately 30 minutes.
- Use the same teams from Activity R, Twist of the Wrist Design Team.
- Activity R – Twist of the Wrist Design Team must be completed before doing this activity.

Focus for Module 3: Get a Grip

Big Ideas
- Use of simple machines
- Form and function
- Geometry of levers and links in mechanical movements

NSE Standards
- Abilities of technological design
- Implementing a proposed design

Performance Tasks For Youth

You will learn about joints and linkage by exploring with chopsticks. You will link (use) two chopsticks together to form a gripper and lift small objects.

You will learn about joints and linkage by exploring various types of end effectors (grippers, tools, etc.). These devices may lift, hold, cut, or squeeze objects and vary in design, depending on the type of object and task to be done.

You will learn about joints and linkage by exploring with chopsticks, pliers, and tongs. You will select a gripper best suited to lift the object.

You will explore how end effectors (grippers) can be assembled and how to build or make more complex things. You will plan and design a gripper using the parts set out for this activity, basing the design on the objective of picking up selected items.

You will build a gripper using the design from Activity P and various parts and supplies.

You will design a way to attach the gripper to your robot arm.

You will fasten the gripper to the robot arm and try it out by grabbing an item and moving it with the air-powered arm.

STL
- Engineering design
- Problem solving
- Apply the design process
- Manufacturing technologies

SET Abilities
- Invent/Implement Solutions
- Draw/Design
- Build/Construct

Life Skills
- Problem Solving
- Cooperation
- Teamwork

Success Indicators

Youth will understand the concept of gripping an object using leverage and pressure.

Youth will be able to determine and sort the type of grippers and tools based on design, construction, or use of the end effectors.

Youth will be able to problem solve and select the best gripper to lift their object.

Youth will be able to sketch a design for a gripper to pick up simple items.

Youth will be able to build a gripper that can successfully grip and hold a selected item.

Youth will complete a sketch/design on merging their robot arms and grippers.

Youth will be able to attach the gripper to their robot arm.

Activity M – Chopsticks

Performance Task For Youth

You will learn about joints and linkage by exploring with chopsticks. You will link (use) two chopsticks together to form a gripper and lift small objects.

Success Indicators

Youth will understand the concept of gripping an object using leverage and pressure.

List of Materials Needed

- Robotics Notebook
- Two straws for each participant
- Two chopsticks for each participant
- A selection of small items to pick up: marbles, balls, ping-pong balls, golf balls, plastic eggs, or wood blocks

Activity Timeline and Getting Ready

- Activity will take approximately 15 minutes.
- Divide youth into small groups of two or three.
- Set parts on the table.

Experiencing

1. Ask participants to use the chopsticks and straws to pick up various items such as marbles, plastic eggs, or pencil erasers.

2. Have participants record in their Robotics Notebook their experiences, what they discovered, and what they learned.

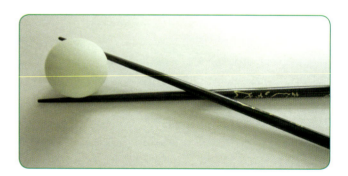

Sharing and Processing

As the facilitator, help guide youth as they question, share, and compare their observations. Before they share with the group, have youth use their Robotics Notebook to record ideas, comments, and notes on the activities they have been doing. You may choose one of the questions below as a prompt. If necessary, use more targeted questions as prompts to get to particular points.

- How did you hold the sticks? What was the fulcrum (pivot)?
- What leverage advantage did you get from the sticks? (More or less pressure?)
- What was difficult about holding onto the items you tried to pick up? What was easy?
- How did the straws work differently than the chopsticks?
- What would make it easier to grab an item?

Generalizing and Applying

- Identify other items that could be used to lift objects.
- How would you modify chopsticks so that you could use them to eat soup? Other food?
- Youth also can apply what they have learned in Activity N.

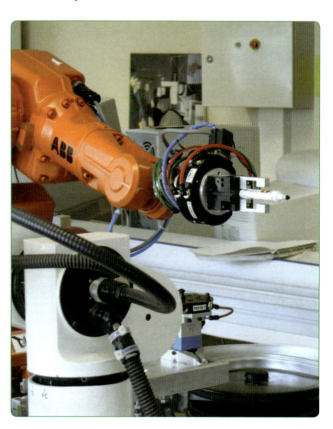

Activity N – Just a Pinch

Performance Task For Youth

You will learn about joints and linkage by exploring various types of end effectors (grippers, tools, etc.). These devices may lift, hold, cut, or squeeze objects. They vary in design depending on the type of object and task to be done.

Success Indicators

Youth will be able to determine and sort the type of grippers and tools based on design, construction, or use of the end effectors.

List of Materials Needed

- Robotics Notebook
- An assortment of pliers, tongs, scissors, and nutcrackers – one for each participant

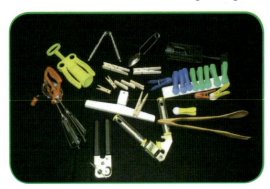

Activity Timeline and Getting Ready

- Activity will take approximately 20 minutes.
- Arrange youth around a table or circle of chairs. If it is a large group, you may set up multiple tables or circles.
- Collect enough grippers (tongs, pliers, etc.) so that each individual will have one.

Safety Note

Some of the grippers can be sharp; others can pinch. Try to select grippers that will reduce the risk of youth hurting themselves or others. Caution the youth not to poke, pinch, or grab others with these grippers.

Experiencing

1. Place enough items, including a variety of tongs and pliers, in the middle of the group so that everyone will be able to select an item.

2. First, ask each participant to pick one item, examine it, and try to make the item work Have each participant share information about the item with the others, such as the type of lever, mechanical advantage, fixed or adjustable, function of item (to move, grasp, cut, or other use).

3. Have members select a different item. With the second item, have the participants share information about the type of fastener (pivot) used in the item. Is it removable? Is it the same type of material? Does it move?

4. Have members sort the items into groups and describe the groupings (materials from which the items are made, the item's use, the type of lever/fulcrum).

5. Have participants record in their Robotics Notebook their experiences, what they discovered, and what they learned.

Sharing and Processing

As the facilitator, help guide youth as they question, share, and compare their observations. Before they share with the group, have youth use their Robotics Notebook to record ideas, comments, and notes on the activities they have been doing. You may choose one of the questions below as a prompt. If necessary, use more targeted questions as prompts to get to particular points.

- How might a robot hand be like any of these items?
- How are these items different from a robot hand?
- Explain how you would re-sort the items. Why?

Generalizing and Applying

- What do you think a robot hand could look like?
- What types of grippers are used at home or school?
- What other simple machines — besides levers — can you find in the items? Screw? Wedge? Wheel/axle? Pulley? Inclined plane?
- Youth also can apply what they have learned in Activity O.

Activity 0 – Hold On

Performance Task For Youth

You will learn about joints and linkage by exploring with chopsticks, pliers, and tongs. You will select a gripper best suited to lift your object.

Success Indicators

Youth will be able to problem-solve and select the best gripper to lift their object.

List of Materials Needed

- Robotics Notebook
- An assortment of pliers, tongs, scissors, nutcrackers, etc. – one for each participant
- A selection of small items to pick up: marbles, balls, ping-pong balls, golf balls, plastic eggs, wood blocks; also, some items that cannot be picked up by the grippers that are available

Activity Timeline and Getting Ready

- Activity will take approximately 20 minutes.
- Divide youth into small groups of two or three.
- Be ready to distribute the items that will be picked up; one item for each group.
- Do not distribute the grippers for the first part of this activity; they are for display only. (During the second part of the activity, they will be selected.)

Experiencing

1. Give each group one item to be gripped and picked up.

2. First, ask each group to discuss the various tongs and pliers and determine which ones they think will work best in gripping and picking up their item without harming or breaking the item. Have them make a list of all the grippers they think would hold/pick up their item and indicate which one would work best. Have the groups share their predictions with the other groups.

3. Have the teams test all of the grippers on their list. Remind them to be careful so they don't harm their "pick up" item. Determine from the testing which grippers worked the best. Were these the same grippers they predicted would work well? Have them share their findings with the other groups.

4. Have participants record in their Robotics Notebook their experiences, what they discovered, and what they learned.

Safety Note

Some of the grippers can be sharp; others can pinch. Try to select grippers that will reduce the risk of youth hurting themselves or others. Caution the youth not to poke, pinch, or grab others with these grippers.

Sharing and Processing

As the facilitator, help guide youth as they question, share, and compare their observations. Before they share with the group, have youth use their Robotics Notebook to record ideas, comments, and notes on the activities they have been doing. You may choose one of the questions below as a prompt. If necessary, use more targeted questions as prompts to get to particular points.

- How is the structure of your hand related to its function?
- Why was the gripper you used made like it was? Was it designed for the function of the part or was it designed for the look and feel of the part?
- How did you predict which gripper would work for your item? What knowledge did you need?
- How did testing the grippers change your predictions?
- How important was it to know how to describe a certain function or movement?

Generalizing and Applying

- Would a gripper for one item be good for another item?
- If you want to pick up something fragile, such as an egg, how would you design the gripper?
- What types of parts would be useful in building a robot gripper?
- Youth can apply what they have learned in Activity P.

Activity P – One for the Gripper Design Team

Performance Task For Youth

You will build a robot arm from your design in Activity H.

Success Indicators

Youth will understand the principles behind arms and movement as demonstrated by their building a robotic arm using levers to pick up and move a weight from one spot to another location.

List of Materials Needed

- Robotics Notebook
- For display only in this activity: items and parts for a gripper such as hinges, bolts, nuts, wood, plastic pieces, craft sticks, syringes, tubing, etc., from the Trunk of Junk
- A selection of small, lightweight items to pick up: ping-pong balls, plastic golf balls, plastic eggs, or toy blocks
- Optional: for reference, the robot arm from Activities I and L

Activity Timeline and Getting Ready

- Activity will take approximately 20 minutes.
- Divide youth into small Design Teams of two or three.
- Display building items for viewing.
- Items to be picked up by the gripper can be available for inspection; teams can be assigned an item or allowed to select an item for their design.

Experiencing

1. First, discuss grippers. Ask participants to think about grippers/holders they have seen, read about, or imagined. Ask them to share their thoughts with the group.

2. Allow the Design Teams to view the parts and supplies available but do not let them touch or hold the items during the design phase.

3. Second, have participants design and draw a gripper that will pick up the item given to the team. Each team may be given a different item to grip, or they may all have the same item for gripping. Since the gripper may be attached to the robot arm they built in Activities I and L, the item to grab should be light, such as a ping-pong ball or plastic golf ball. Also, the gripper should be made small enough to be attached to the robot arm they built in Activities I and L.

4. Have the Design Teams use their Robotics Notebook to sketch their gripper designs.

5. Have the teams share their design for a gripper.

Sharing and Processing

As the facilitator, help guide youth as they question, share, and compare their observations. Before they share with the group, have youth reflect on the activity in their Robotics Notebook. You may choose one of the questions below as a prompt. If necessary, use more targeted questions as prompts to get to particular points.

- What are some ways to assemble the parts of a gripper?
- What types of joints and linkages would make good grippers?
- What other parts or supplies might make it easier to build this gripper?

Generalizing and Applying

- What uses would you have for a gripper at home? At school? Other places?
- What are some things that are held — or should be held — by a gripper?
- Youth can apply their designs by building their gripper in Activity Q.

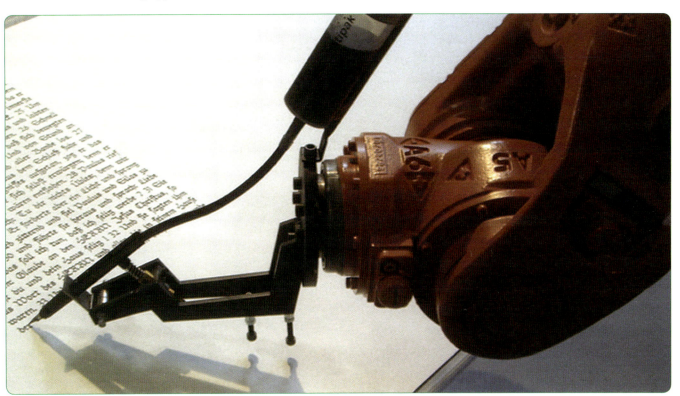

Activity Q – One for the Gripper Build Team

Performance Task For Youth

You will build a gripper using the design from Activity P and various parts and supplies.

Success Indicators

Youth are able to build a gripper that can successfully grip and hold a selected item.

List of Materials Needed

- Robotics Notebook
- Items for a gripper, including assorted parts, fasteners, hinges, bolts, nuts, wood, plastic pieces, craft sticks, syringes, tubing, etc., from the Trunk of Junk
- A selection of small items to pick up: ping-pong balls, plastic golf balls, eggs, or toy blocks
- Simple tools – saws, drills, and other tools
- Optional: for reference, robot arm from Activities I and L

Activity Timeline and Getting Ready

- Activity will take approximately 50 minutes.
- Use the same teams from Activity P – One for the Gripper Design Team.
- Make parts and supplies available for building.
- Provide a work area for tools and assembly.
- Activity P – One for the Gripper Design Team must be completed before doing this activity.

Experiencing

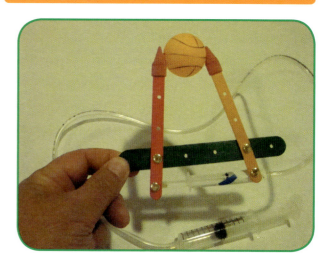

1. Use the ideas from the Design Team to build a gripper from the parts and fasteners.
2. Give each team one item (plastic golf ball, block, or other item) to be grabbed and held by the gripper.
3. Provide an assortment of parts and fasteners to use in building the gripper.
4. Ask the teams to build a gripper powered by a syringe so they can pick up the item they have been given.
5. When the grippers are complete, have the teams demonstrate how they are able to pick up the items for which the grippers were designed.
6. Ask the teams to try to pick up the other items. Do their grippers work as well with the other items?
7. The Build Teams should record actions and modifications in their Robotics Notebook.

Activity R – Twist of the Wrist Design Team

Youth will combine the work of their robot arm, power source, and gripper into one robot! Activities R and S require the completion of Activities I, L, and Q.

Performance Task For Youth

You will design a way to attach the gripper to your robot arm.

Success Indicators

Youth will complete sketches/designs on merging their robot arms and grippers.

List of Materials Needed

- Robotics Notebook
- A robot arm (see Activities I and L)
- A gripper (see Activity Q)
- For display only in this activity: items and parts for attaching, fastening, and combining, such as hinges, small bolts, nuts, plastic pieces, craft sticks, etc., from the Trunk of Junk

Activity Timeline and Getting Ready

- Activity will take approximately 20 minutes.
- Divide youth into small Design Teams of two or three.

Experiencing

1. Participants will plan and design how to add their gripper from Activity Q to the robot arm they made in Activities I and L.
2. Review and discuss the three degrees of freedom for robot wrist movements: yaw, bend, and rotation. (See page 35.)
3. Encourage attachment designs that allow at least one degree of freedom at the wrist.
4. Have the Design Teams use their Robotics Notebooks to sketch their attachment designs.
5. Have the teams share their design for a gripper.

Sharing and Processing

As the facilitator, help guide youth as they question, share, and compare their observations. Before they share with the group, have youth use their Robotics Notebook to record ideas, comments, and notes on the activities they have been doing. You may choose one of the questions below as a prompt. If necessary, use more targeted questions as prompts to get to particular points.

- What type of attachment are you using?
- Which of the three wrist movements will it allow, if any?

Generalizing and Applying

- What items at school would benefit from having one or more robotic wrist movements? What about items at home?
- What already has some of these movements?
- Apply your design by attaching it to a robot arm in Activity S.

Activity S –Twist of the Wrist Build Team

Performance Task For Youth

You will fasten the gripper to the robot arm and try it out by grabbing an item and moving it with the air-powered arm.

Success Indicators

Youth will be able to attach the gripper to their robot arm.

List of Materials Needed

- Robotics Notebook
- A robot arm (see Activities I and L)
- A gripper (see Activity Q)
- Items and parts for attachment, fastening, and combining, such as hinges, small bolts, nuts, plastic pieces, craft sticks, etc., from the Trunk of Junk
- Simple tools – saws, drills, and other tools

Activity Timeline and Getting Ready

- Activity will take approximately 30 minutes.
- Use the same teams from Activity R, Twist of the Wrist Design Team.
- Activity R, Twist of the Wrist Design Team, must be completed before doing this activity.

Experiencing

1. Have the Design Teams use their ideas to attach the grippers to the arms.
2. After the grippers are attached, ask the teams to use them to pick up the assigned items and move them to another location using the air power from syringes.
3. The Build Teams should record actions and modifications in their Robotics Notebooks.

Sharing and Processing

As the facilitator, help guide youth as they question, share, and compare their observations. Before they share with the group, have youth reflect on the activity in their Robotics Notebook. You may choose one of the questions below as a prompt. If necessary, use more targeted questions as prompts to get to particular points.

- What are some ways that worked to attach the gripper?
- How did the differences in arm size, leverage, or gripper size affect the attachment and use of the total robot arm and gripper?
- How could you add more wrist movements? What problems would have to be overcome to get all six movements of the robot arm and wrist? Refer to the description of the six movements on page 51.

Generalizing and Applying

- What problems do engineers have to overcome in designing tools and equipment like automated garbage trucks, tractor backhoes, construction cranes, or carnival rides? Other types of arms and grippers?

What's Next?

You have completed learning about robotic arm systems, designing your own arm, and using technology tools to build and assemble an arm powered by air. You have been a scientist in learning, an engineer in designing, and a technician in building.

In Junk Drawer Robotics Level 2, Robots on the Move, you can continue to learn, to do, and to make robots that walk, roll, and dive underwater. You'll use electricity as a power source and gears for movement.

You also can check out the other 4-H Robotics tracks on Virtual Robotics and Platform Robotics. In these tracks, you can design and build robots online or use kits to build and program robots using a computer.

To learn more about these tracks, visit the website at: *www.4-H.org/curriculum/robotics*.

Glossary

Module 1 – Parts Is Parts

- **Build** – formed by fitting or joining components together
- **Construct** – put together by combining materials and parts
- **Design** – make or work out a plan, often in graphic form, to invent
- **Fabricate** – to make, build, or construct from raw material by assembling parts or manufacturing
- **Form** – the look, the feel, or the beauty of an object
- **Function** – the use or need filled by an object
- **Make** – created by a manufacturing process
- **Produce** – create or manufacture a manmade product
- **Manufacture** – the change of raw materials into finished goods

Module 2 – In Arm's Reach

- **Balance** – an ability to maintain the center of gravity
- **3-D space** – movement in all three directions: up and down, left to right, and backward and forward
- **Arm** – robot manipulator connected by joints allowing for motion
- **Pneumatics** – using pressurized gas to provide mechanical motion
- **Hydraulics** – machines that use fluid power to do work. Fluid is controlled by valves and distributed through hoses
- **Cartesian coordinates** – a system to indicate the location of a point on a surface using two axis lines (X,Y) that can locate a point in three-dimensional space by adding a third axis line (Z)
- **Torque** – how hard something is rotated or twisted

Module 3 – Get a Grip

- **Continuous path motion** – programming a robotic arm in very fine movements so that the hand can follow a specific path, such as next to a curved, flowing surface
- **Electro-magnet** – a piece of iron wrapped in a coil of electrical wire acting with magnetic force when electricity is turned on in the coil
- **End effector (hand, gripper, or tool)** – device at the end of an arm designed to do something with a robot's surroundings
- **Joint** – location at which two or more levers or parts contact, allowing movement between those parts
- **Linkage** – rods, levers, or parts connected with joints to allow motion in one place to be transmitted to another location
- **Mechanical** – having to do with mechanics and being done by a machine as in control of motion or power
- **Point to point motion** – programming a robotic arm with the main focus on the location of the beginning and ending points
- **Leverage** – using a lever and pivot point to adjust the force in an application
- **Pressure** – the amount of force applied to an area